AF710351

R & D MANAGEMENT SYSTEMS IN JAPANESE INDUSTRY

R & D MANAGEMENT SYSTEMS IN JAPANESE INDUSTRY

edited by

Hajime ETO
The University of Tsukuba
Ibaraki, Japan

and

Konomu MATSUI
Rikkyo University
Tokyo, Japan

1984

NORTH-HOLLAND – AMSTERDAM · NEW YORK · OXFORD

© ELSEVIER SCIENCE PUBLISHERS B.V., 1984

All rights reserved. No part of this publication may be reproduced, stored in a retrieval system, or transmitted, in any form or by any means, electronic, mechanical, photocopying, recording or otherwise, without the prior permission of the copyright owner.

ISBN: 0 444 86808 9

Published by:
ELSEVIER SCIENCE PUBLISHERS B.V.
P.O. Box 1991
1000 BZ Amsterdam
The Netherlands

Sole distributors for the U.S.A. and Canada:
ELSEVIER SCIENCE PUBLISHING COMPANY, INC.
52 Vanderbilt Avenue
New York, N.Y. 10017
U.S.A.

HD
45
.R2
1984

Library of Congress Cataloging in Publication Data

Main entry under title:

R & D management system in Japanese industry.

1. Technological innovations--Japan--Addresses, essays, lectures. 2. Research, Industrial--Japan--Addresses, essays, lectures. I. Etō, Hajime, 1935- II. Matsui, Konomu, 1931- . III. Title: R and D management system in Japanese industry.
HD45.R2 1984 658.5'7'0952 83-20489
ISBN 0-444-86808-9

PRINTED IN THE NETHERLANDS

PREFACE

The increasing interest in Japanese Innovations leads to an interest in R&D management system and strategy in Japanese industry. As Japanese management differs from Western one and Japanese innovations differ from Western ones, R&D management system and strategy formation process may be quite different between Japan and West. Recognizing this difference will surely contribute to the methodalogical progress in the related disciplines and interdisciplinary approaches.

No particular part of this book makes a comparative study between Japan and West or a unification of Japanese and Western approaches explicitly. But readers will find relavant informations for that everywhere in this book.

This book is composed mainly of reports of real activities in industry with a few exceptions of new proposal, observation and review. The affiliations of authors reveal backgrounds on which the works were done. But the views stated are those of authors and do not necessarily reflect the policies of these affiliated organizations.

Chapter I: "Needs-Oriented Structural Analysis for Fifth Generation Computer Systems" reports a technical work in the top computer maker in cooperation with government-related organizations. Readers will see how a computer development strategy is formed via informal interactions between the government and industry and what needs motivate individual scientists to develop and apply a management tool in such an informally organized "project".

Chapter II: "Toward a Policy Guidance System for Complex Innovation" conceptualizes experiences in the top technological giant and proposes a new system on the basis of the experiences.

Chapter III: "Problem of Present Evaluation of Public Investment and Proposal of New Evaluation of R&D Investment as Reflationary

Policy" casts a doubt on the prevailing idea of public investment among economists and financiers and proposes a new criterion for investment. This chapter provides an example of individual's initiative in generating alternative policies in Japanese offices.

Chapter IV: "On Japanese Creativity" is an observation on Japanese creativity by an western expert in Japanese QC (quality control). The idea of Japanese QC movement is new even to Japanese people. This chapter outlines a creativeness aspect of Japanese QC and tries to foresee its future development.

Chapter V: "A Framework for Investigations of Information Technology-Based Innovation" displays how a concept is formed for a corporate innovation policy by integrating many ideas proposed outside a company. The reference list at the end of this chapter provides a profile of a trend of the Japanese media on innovations.

Chapter VI: "Recent Trends in R&D and Patents - A Quantitative Appraisal -" is only one chapter of economic approach in this book. Applying the regression analysis, this chapter compares the recent R&D productivity of Japan with others.

Chapter VII: "Behaviour of Japanese R&D Organizations - Conflict and its Resolution by Informal Mechanisms -" gives a historical sketch of science and technology policy to provide explanatory informations on backgrounds of R&D management systems. Focus is on the development of conflict between organizations and its resolution. This chapter recalls attention to a foregotton aspect of conflict behind Japanese consensus.

Chapter VIII: "Soft Science in Japan - Ten Year's Experience in Mitsubishi Electric Corporation -" reports a case of development of a Japanese-made new science in a traditional company. Comparing it with Western counterparts, readers can unveil characteristics of a trend of Japanese R&D management systems in this decade. Readers will notice that such a future-oriented group grows in such a large and traditional company rather than in a small and new company. This chapter tells that the Japanese concept of venture business is different from the Western one.

Some of chapters in this book are based on presentations at Technological Forecasting Symposium held annually by JATES (Japan Techno-Economic Society) since 1972. The indirect contribution of this symposium to this book, or generally to systems approach to R&D in Japan is sincerely appreciated.

H.Eto

CONTENTS

R & D Management Systems in Japanese Industry
H. Eto and K. Matsui (editors)
© Elsevier Science Publishers B.V. (North-Holland), 1984

Chapter I NEEDS-ORIENTED STRUCTURAL ANALYSIS FOR FIFTH GENERATION COMPUTER SYSTEMS

MITSUHIKO TODA and KOZO SUGIYAMA

Policy Science Unit
International Institute for Advanced Study of
Social Information Science, Fujitsu Limited
140 Miyamoto, Numazu, Shizuoka 410-03, Japan

A needs-oriented analysis is presented for research and development of application technologies for fifth generation computer systems. A brief description is given regarding a needs analysis that identifies application fields of the computer systems. A method for structural analysis and strategy analysis was originated for the needs-oriented analysis in order to investigate combinatorial relations of the needs and technologies. Important application technologies of the fifth generation computer systems are explicitly identified by applying the method. Strategies for their developments are obtained by the analysis.

1 Introduction

National efforts have been made in Japan since 1982 to develop so called 'fifth generation computer systems' [1]. Research and development (R & D) of the computer systems is one of multiple undertakings to attain the proposed national goals for the '80s [2] ; (1) contribution to the international community, (2) overcoming limitations of resources and energy, and (3) the co-existence of social dynamism and the improved quality and comfort of life. A perspective toward a technology-based nation has also been proposed as one of the strategic means to attain these goals. One objective of technological developments in the perspective is to develop knowledge-intensive technologies. Technologies for the fifth generation computers, which will enable us to construct versatile knowledge information processing systems, will play an important role in developing knowledge-intensive technologies in various related fields.

The proposal also points out that a change in technology policy is essential for Japan to switch over from improvement and application of imported technologies to the research and development of original technologies. Emphasis is placed upon the following three points in systematically pursuing technology policies in response to this new phase.

(1) development of original and creative technologies

(2) systematic promotion of technological developments

(3) increased allocation of R & D funds

Concerning item (2) the proposal insists that 'technological developments must be promoted by presenting a "Long-term Vision for Technological Development," which identifies the priority goals for technological developments, as well as systems for development and funding' [2]. The R & D of fifth generation computer systems (FGCS) is a representative national project that has been launched and promoted through a policy based on these three points.

The purpose of this paper is to present results of a needs-oriented analysis for FGCS. The analysis is based on data that identify anticipated social needs and relate them to application technologies of FGCS, and a method for structural analysis is applied to yield holistic relational structures of those needs and relevant technologies. Important fields of application and promising computer technologies are identified by observing the structures of their relations illustrated by hierarchical graphs and by the use of priority measures. Suggestions on strategies for developing the technologies will be presented based on their priorities.

The input data for our analysis were obtained by a study of social environments for FGCS [3]. The study focuses on fields of application for emerging information technologies, particularly FGCS which will play an important role. This study layed the foundation for generating the ten year R & D plans for the Institute for New Generation Computer Technology [4], which aim at developing basic computer techniques for knowledge information processing systems. In this paper we are concerned with all the fields of application considered in the study, and we present a systems analysis on the relations of application technologies of the basic computer techniques and relevant information technologies. Thus we identify important application technologies and suggest strategies for their developments.

Decision making for technological developments requires us to take various aspects of the developments into consideration. One of the important aspects is an evaluation of their potential future growth; their projected success in meeting future social needs. However, it is difficult to forecast quantitatively, as one can in economics, which social needs will demand technologies. Such forecasting is particularly difficult in the case of the above mentioned innovative and original technologies. The innovative nature of the technologies makes forecasting of social needs difficult due to a lack of analogical inference based on similar technologies presently in use, because as innovative technologies diffuse in societies they often generate unexpected demands. Another reason of the difficulty is the long time required for research and development of innovative technologies; hence, a great deal of uncertainty arises in the long range forecasting of social environments. Therefore, long range planning of R & D for innovative technologies often needs to be based on comparatively less quantified data which are subjectively assessed by experts and decision makers. The development of fifth generation computers, projected to be in commercial use in the 1990s, is an example of such innovative technological developments.

The method utilized for the needs-oriented analysis consists of Visual Q-Analysis (VQA) for structural study and strategy analysis based on VQA. VQA is appropriate for the analysis, because it can generate a holistic relational structure of needs and technologies based on comparatively less quantified relational data obtained from Ref.[3]. Diversified social needs anticipated for FGCS require various intelligent functions from the computers. On the one hand an integrated technology for a need comprises various element information technologies, and on the other hand an element information technology is required for various integrated technologies. VQA is particularly suitable for analyzing combinatorial relations of these technologies.

This paper consists of five sections. In Section 2, we briefly describe Japanese efforts to research snd develop FGCS, and the analysis which identifies needs for the computers. In Section 3, we present a brief description of the method originated for this study. Section 4, the main section of this paper, presents findings of this study obtained from structural models of the needs and relevant technologies for the computer systems. Strategy analysis is also described by making use of the models. Section 5 concludes the paper with remarks regarding further research.

2 Japanese R & D efforts for fifth generation computer systems

Concerted efforts preparatory for the development of FGCS began in 1979. The Japan Information Processing Development Center (JIPDEC) conducted research during the three fiscal years 1979-1981 previous to the launching of the national project. A summary of the first two years' study was reported at the International Conference on Fifth Generation Computer Systems in 1981 [1]. The JIPDEC research committee for the study was organized into three subcommittees ; (1) social environment (first year) and system integration (second year) subcommittee, (2) basic theory subcommittee, and (3) architecture subcommittee. From this broad perspective the committee clarified requirements for FGCS and subjects for R & D. JIPDEC is a nonprofit organization supported by the Japanese Government and information industries. Therefore, the studies served to target national efforts of Japan to develop FGCS which will promote knowledge-intensive technologies and a technology-based nation.

The JIPDEC study resulted in the Institute for New Generation Computer Technology (ICOT), which was established in April 1982 to develop basic computer techniques for FGCS. ICOT is a nonprofit research organization, and is supported by the Japanese Government, academic institutions and computer manufacturers [5]. ICOT adopted its ten-year R & D plans based on the JIPDEC study, although ICOT was established as an organization independent of JIPDEC.

2.1 Fifth generation computer systems

FGCS conceptualized by ICOT is a versatile knowledge information processing system with emphasis on a modeling (software) system, and its functions are classified as follows [4].

(1) problem-solving and inference

(2) knowledge-base management

(3) intelligent interface

These functions are realized by the modeling (software) system consisting of an intelligent programming system, a knowledge-base system and intelligent interface systems for communication with human users employing speech, natural languages and pictures etc. The software system will run on a hardware system consisting of a problem solving and inference machine, and a knowledge machine, in which a kernel logic programming language will play a key role as their interface.

2.2 Identification of needs and application technologies

The JIPDEC social environment subcommittee investigated requirements for FGCS to cope with social difficulties which Japan will face in the 1990s in an effort to construct a desirable society [3]. The subcommittee was composed of a group of six experts ; two specialists each on technological innovations, in information industries, and users of information systems. These experts analyzed future states of the nation for the purpose of specifying social conditions that should be considered in determining the target of national efforts to develop FGCS. They analyzed basic statistical data which indicate past trends in Japan and forecast by extrapolation through the 1990s. In addition to the data analysis specialists' opinions were collected about the social environments, thereby arriving at five social problems (these were called 'social bottlenecks') anticipated in the 1990s. Information technologies were considered as one of the technological means to eliminate the bottlenecks together with socio-political means. Although information technologies may play a subordinate role for this purpose, the technologies will be none the less important as a means to help alleviate the bottlenecks and to foster an information/knowledge intensive society. In order to relate the social bottlenecks, which cover general areas of social problems, to information technologies, the experts expanded them to generate a hierarchy of social requirements, where the requirements become consecutively more specific. The requirements in the third (bottom) level of the hierarchy were so selected that they are specific enough to yield necessary information technologies.

The five bottlenecks (B_1-B_5) specified by the subcommittee are

- B_1 : improvement in the fields of low productivity such as office work, engineering design, agriculture and fishery, medicine, education, public services, and government
- B_2 : internationalization of Japan through a transferral of its knowledge to the world and a maintenance of competitive power in international trade by overcoming shortages in energy and natural resources
- B_3 : coping with problems generated by structural changes in Japanese society such as an increase in the population of the aged, changes in attitudes of younger generations toward labor, changes in industrial structures, etc.
- B_4 : increased efficiency of social systems and reduction of communication gaps through computerization (promoting an information society)
- B_5 : effective information systems for individuals (fostering an information effective life)

The subcommittee expanded these five bottlenecks to obtain 30 specific social requirements, which are called social needs for information technologies. The hierarchy of social requirements obtained by the break down is shown in Figure 1. The needs denoted by N_i in Figure 1 are described in Table 1, which also lists 37 information technologies (S_j) that will be required to meet those needs. The needs in Figure 1 and Table 1 were not meant to be exhaustive, but they were selected by the subcommittee as representative application fields for information technologies. The subcommittee argues that the needs in Table 1 are already recognized as application technologies of general interest. Table 2 summarizes requirement relations assessed by the subcommittee and specifies the information technologies that were assessed necessary for each need. In Table 2 the notation $N_i = \langle S_j, \ldots, S_k \rangle$ specifies that technologies $S_j, \ldots, S_k$ are required for need N_i. The extensive scope of the technologies considered in Tables 1 and 2 contains various information technologies related to FGCS as well as application technologies of FGCS.

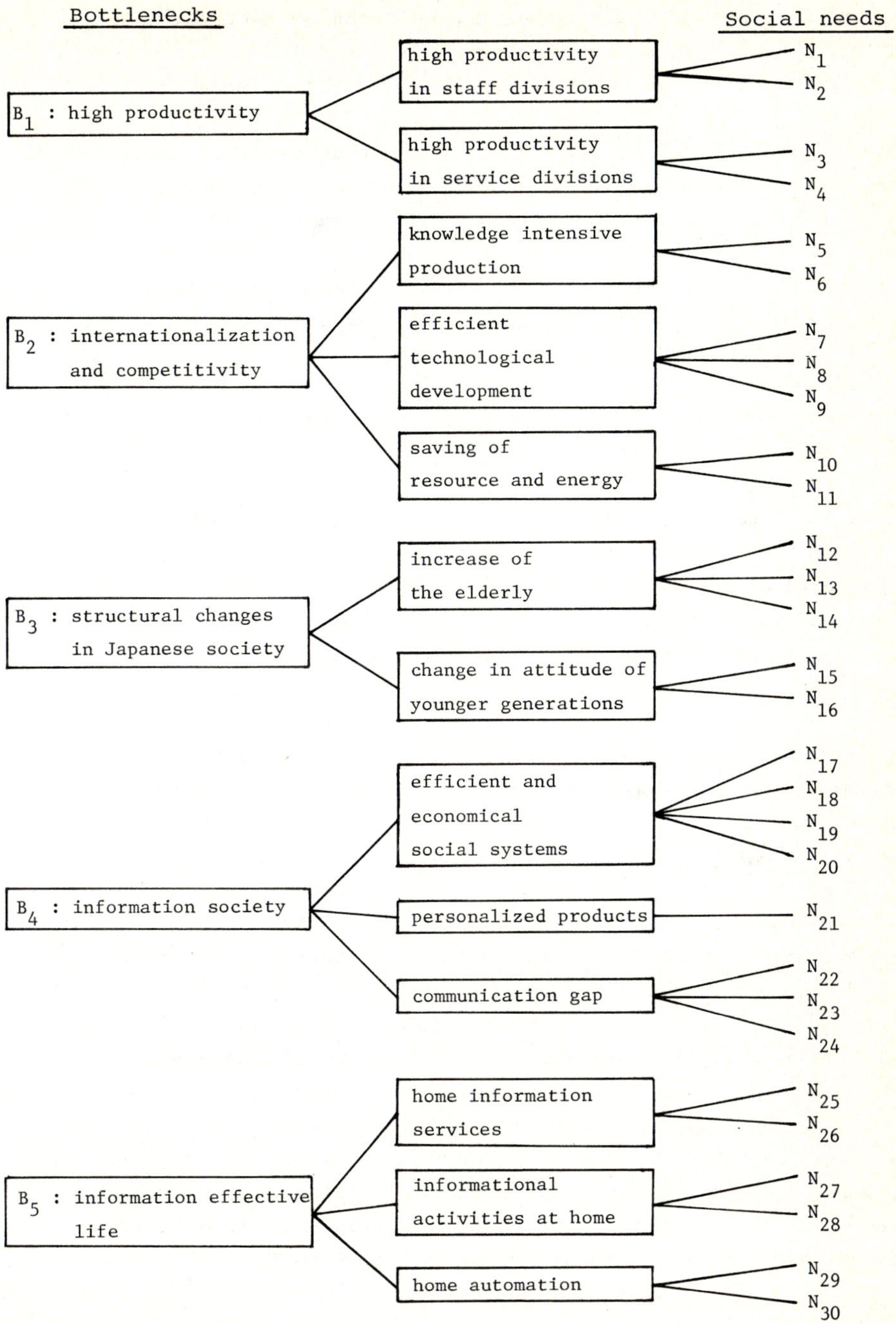

Figure 1 Hierarchy of social requirements

Table 1 Needs(N_i) and information technologies(S_j) identified

N_1 : decision support systems for planning and staff activities
N_2 : office automation for routine office work
N_3 : office automation for real-time clerical services
N_4 : very large data bases for service sectors
N_5 : intelligent machines : robots & CAM(computer aided manufacturing)
N_6 : diversification of custom made products
N_7 : IR (information retrieval) for technological information
N_8 : design automation
N_9 : laboratory automation
N_{10}: control systems for saving energy
N_{11}: conversion to energy/resource saving industrial structure with new information industries
N_{12}: medical electronics for medical offices
N_{13}: education for senior citizens
N_{14}: health care industries
N_{15}: personal computers for young generations
N_{16}: robots for dirty work
N_{17}: medical service systems
N_{18}: CAI (computer aided instruction) for personal education
N_{19}: disaster prevention IR systems
N_{20}: crime prevention IR systems
N_{21}: personal computers for custom products
N_{22}: computer network for regional information services
N_{23}: CAI for communication among generations
N_{24}: machine translation of languages
N_{25}: integrated reservation systems accesible at home
N_{26}: telemail and teletext
N_{27}: CAI for learning at home
N_{28}: hobby computers
N_{29}: program-controlled electric appliances
N_{30}: home automation

S_1 : supercomputers
S_2 : parallel processing architecture
S_3 : highly-integrated microcomputers
S_4 : very large storage technology
S_5 : packaged software
S_6 : computer games
S_7 : computer arts
S_8 : simulation technology
S_9 : information processing services
S_{10}: software development services
S_{11}: TSS services
S_{12}: facility management systems
S_{13}: computer network services
S_{14}: computer lease
S_{15}: digital network
S_{16}: picture and image processing
S_{17}: video technology
S_{18}: distributed processing
S_{19}: distributed databases
S_{20}: document databases
S_{21}: engineering databases
S_{22}: protection of data security
S_{23}: associative retrieval
S_{24}: data services
S_{25}: picture processing
S_{26}: picture and image processing
S_{27}: character and picture recognition
S_{28}: optical input/output processing
S_{29}: Japanese language processing
S_{30}: natural language processing
S_{31}: intelligent terminals
S_{32}: terminals for counter services
S_{33}: health care equipments
S_{34}: sensor bases
S_{35}: built-in microcomputers
S_{36}: databases
S_{37}: man-machine interface

Table 2 Information technologies required for each need

$N_1 = \langle S_{23}, S_{25}, S_{29}, S_{36} \rangle$

$N_2 = \langle S_{20}, S_{23}, S_{25}, S_{28}, S_{29} \rangle$

$N_3 = \langle S_{18}, S_{25}, S_{27}, S_{29}, S_{31}, S_{32} \rangle$

$N_4 = \langle S_4, S_{19}, S_{22}, S_{27}, S_{29} \rangle$

$N_5 = \langle S_5, S_{35} \rangle$

$N_6 = \langle S_2, S_{21}, S_{26} \rangle$

$N_7 = \langle S_{23}, S_{30}, S_{36} \rangle$

$N_8 = \langle S_2, S_{21}, S_{26} \rangle$

$N_9 = \langle S_1, S_{34} \rangle$

$N_{10} = \langle S_{34}, S_{35} \rangle$

$N_{11} = \langle S_9, S_{10}, S_{11}, S_{12}, S_{13}, S_{14}, S_{24} \rangle$

$N_{12} = \langle S_{22}, S_{23}, S_{26}, S_{31}, S_{34}, S_{36} \rangle$

$N_{13} = \langle S_{25}, S_{29}, S_{35}, S_{36}, S_{37} \rangle$

$N_{14} = \langle S_{33} \rangle$

$N_{15} = \langle S_3, S_5, S_{35} \rangle$

$N_{16} = \langle S_{34}, S_{35}, S_{36} \rangle$

$N_{17} = \langle S_{22}, S_{23}, S_{26}, S_{34}, S_{35}, S_{36} \rangle$

$N_{18} = \langle S_{25}, S_{29}, S_{35}, S_{37} \rangle$

$N_{19} = \langle S_8, S_{34} \rangle$

$N_{20} = \langle S_{22}, S_{27}, S_{36} \rangle$

$N_{21} = \langle S_3, S_5, S_{35} \rangle$

$N_{22} = \langle S_{15}, S_{16} \rangle$

$N_{23} = \langle S_6, S_7, S_{35} \rangle$

$N_{24} = \langle S_1, S_4 \rangle$

$N_{25} = \langle S_{15}, S_{19} \rangle$

$N_{26} = \langle S_{15}, S_{31} \rangle$

$N_{27} = \langle S_{17}, S_{31} \rangle$

$N_{28} = \langle S_5, S_{26}, S_{35} \rangle$

$N_{29} = \langle S_{35} \rangle$

$N_{30} = \langle S_{34}, S_{35} \rangle$

The above analysis of the social environments for FGCS was utilized to generate the ten year R & D plans of ICOT, which aim at developing basic computer techniques for FGCS : hardware, base software and basic applications system software (experimental machine translation systems and consultation systems)[4].

In this paper we are concerned with applications of FGCS, and we analyze structural relations of all the needs and technologies in Table 1, so that we can systematically understand application fields for FGCS. The authors do not completely agree with the subcommittee on the needs and technologies identified in Table 1. For example, certain needs for international cooperation, e.g. international information network systems should be identified regarding bottleneck B_2. That is, we believe that Japan's contribution to the international community through cooperation is vital for the nation. Despite the details on which we do not agree, we believe that Table 1 provides a reasonable set of social needs and technologies necessary to analyze FGCS application technologies, especially when we take into account the difficulties involved in identifying future national and social needs for information technologies.

3 Method : Visual Q-Analysis and strategy analysis for R & D

3.1 Integrated information technologies for social needs

Table 2 indicates that each social need N_i requires a combination of information technologies S_j and system integration of these. Integration is necessary to meet specific aspects of the need that are not readily attained by the simple combination of the technologies. The outcome of this system integration may be called an integrated technology I_i to meet need N_i, and in contrast to that each information technology S_j may be called an element technology. Each integrated technology I_i, therefore, materializes specific functions required for need N_i by combining element technologies and integrating them. Table 2 also indicates that each element technology S_j is required for many social needs N_i, and hence is utilized to develop many integrated technologies I_i. This relation among needs N_i, integrated technologies I_i and element technologies S_j can be illustrated as in Figure 2. In the sequel, we will discuss needs N_i and integrated technologies I_i interchangeably when we discuss their relations regarding sharing aspects of the element technologies.

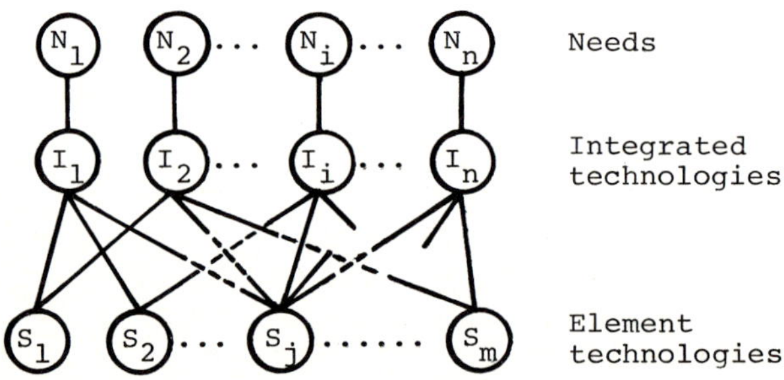

Figure 2 Relations of needs, integrated technologies and element technologies

The concept of integrated technologies is by no means new, but it needs to be emphasized when we consider information technologies. Since information technologies consist more of software elements and hence are more flexible than other technologies, more varieties of integrated technologies can be constructed by integrating element technologies to meet a wide range of social needs.

3.2 Visual Q-Analysis

Visual Q-Analysis is a method for structural analysis of a system based on relational data of the elements of the system [6,7]. Visual Q-Analysis (VQA) owes its basic concepts to Q-Analysis [8] and presents visual models of the system in the form of hierarchical graphs. Required input data to construct the VQA models are pairwise relations of elements of the system in the form $Y_i r X_j$, which denotes that element Y_i is related to element X_j. Here the set of elements $\{Y_i\}$ and that of $\{X_j\}$ are assumed mutually disjoint.

This data requirement is well suited to analyze relations of needs N_i (or integrated technologies I_i for the needs) and those of element information technologies S_j starting from the realtional data in Table 2. The data in the table can be expressed by the above relations ; for example, the technology requirement data for need N_1 yield $N_1 r S_{23}$, $N_1 r S_{25}$, $N_1 r S_{29}$, and $N_1 r S_{36}$. Here the set of needs $\{N_i\}$ and that of technologies $\{S_j\}$ represent disjoint aspects of the scope of the analysis, application fields of FGCS.

We will briefly describe procedures and hierarchical models of VQA in the following. Readers are referred to Ref.[6,7] for methodological details.

For the purpose of illustrating procedures of VQA in the context of R & D let us consider the simple relational data expressed by the incidence matrix in Table 3.

	X_1	X_2	X_3	X_4	X_5	X_6	X_7	X_8	X_9	X_{10}	X_{11}	X_{12}
Y_1	1	1	1	1	1	1	0	0	0	0	0	0
Y_2	0	1	1	0	0	0	1	1	1	0	0	0
Y_3	0	0	0	0	0	0	0	0	0	1	0	0
Y_4	0	0	0	0	0	0	1	0	0	0	1	1
Y_5	1	0	1	0	0	0	1	0	0	0	0	0

Table 3 Incidence matrix

It is assumed that row i and column j of the matrix correspond to need Y_i and element technology X_j respectively, and that its (i,j) component specifies whether or not need Y_i requires technology X_j by

$$(i,j)\ \text{component} = \begin{cases} 1 : \text{need } Y_i \text{ requires technology } X_j \\ 0 : \text{need } Y_i \text{ does not require technology } X_j \end{cases} \qquad (1)$$

An application of VQA to Table 3 yields two kinds of hierarchical models ; the F-Hierarchy in Figure 3 and the Q-Hierarchy in Figure 4, which illustrate combinatorial relations of needs and relevant technologies.

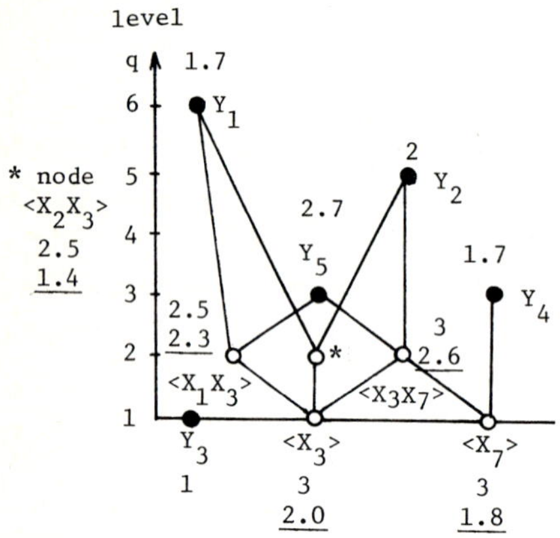

Number attached to a node indicates concentricity

Number attached to a node indicates utility indicator

Figure 3 F-Hierarchy

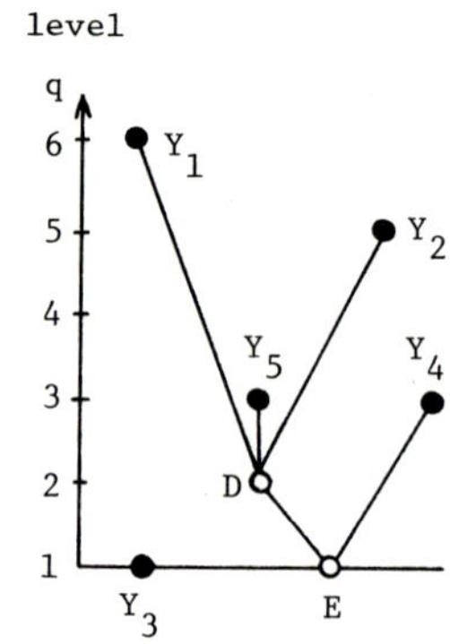

Figure 4 Q-Hierarchy

In the F-Hierarchy the node denoted by ● Y_i represents need Y_i and ● nodes (need nodes) are connected by edges to the nodes denoted by **o**, which represent technologies X_j and their combinations that are commonly required for more than one need and hence are important. These nodes are vertically located according to the number of technologies they represent, which is indicated by level q on the ordinate of the F-Hierarchy. Node ● Y_i is located at the level corresponding to the number of element technologies need Y_i requires.

The F-Hierarchy in Figure 3 illustrates an important connectivity structure among the needs. Node $\circ \langle X_1\ X_3 \rangle$ connected to nodes $\bullet Y_1$ and $\bullet Y_5$ represents that technologies X_1 and X_3 are commonly required for these two needs. Node $\circ \langle X_3 \rangle$ connected to nodes $\circ \langle X_1\ X_3 \rangle$, $\circ \langle X_2\ X_3 \rangle$ and $\circ \langle X_3\ X_7 \rangle$ shows that technology X_3 is shared by these combinations of technologies, which in turn are shared by needs Y_1, Y_2 and Y_5, and hence we can visually recognize that technology X_3 is promising in the sense that it will be utilized in more integrated technologies meeting social needs than other technologies. The technologies required for only one need are not illustrated in Figure 3.

The Q-Hierarchy in Figure 4 illustrates a simplified connectivity structure of the F-Hierarchy, and visually presents results of an application of Q-Analysis to Table 3. The $\circ$ nodes (called connecting nodes) in the Q-Hierarchy simply indicate proximity of needs with respect to the number of common technologies the needs require. The higher the level of connecting nodes of needs, the more closely they are connected. For example, the node $\circ$ D connects needs Y_1, Y_2 and Y_5, and is located at level 2. This implies that each of the three needs requires the same two technologies that are required by some other need of the three. The node $\circ$ D, therefore, may be interpreted as a condensation of the nodes $\circ \langle X_1\ X_3 \rangle$, $\circ \langle X_2\ X_3 \rangle$ and $\circ \langle X_3\ X_7 \rangle$ in the F-Hierarchy. In general, a Q-Hierarchy comprises a set of trees (tree graphs), and illustrates a hierarchical clustering structure of Q-Analysis, hence is a kind of relevance tree.

The SKETCH system [9] is useful for drawing F-Hierarchies and Q-Hierarchies in visually well-arranged hierarchical graphs, so that decision makers can make the best use of these hierarchical models for planning. The SKETCH system enables computers to draw hierarchical graphs automatically for easy visual understanding of structures, by properly arranging horizontal positions of nodes and drawing edges so that connectivity of nodes is easily recognized due to less crossings of edges and nodes of close connection are located close to each other etc. The hierarchical graphs presented in Section 4 were automatically drawn by this SKETCH system.

3.3 Strategy analysis

The hierarchical models obtained by VQA, especially F-Hierarchy, provide solid models for considering strategies of R & D. There are structural measures that are computed from relational data like Table 3 and indicate proximity of needs and technologies. These measures are useful for finding promising technologies and their combinations, and assist us in considering strategies on the visual models generated by VQA. We describe the measures utilizing the example of Table 3.

(a) Concentricity [7] : Concentricity $Con(Y_i)$ of need Y_i is defined by

$$Con(Y_i) = \frac{\Sigma(\text{number of needs requiring technology } X_j)}{\text{number of technologies } Y_i \text{ requires}} \quad (2)$$

where the summation for the numerator is computed for all the element technologies X_j that are required for need Y_i.

Concentricity $Con(Y_i)$ indicates the effectiveness of developing element technologies for need Y_i in terms of how much the technologies X_j developed for Y_i will be utilized for other needs in the scope of analysis. The larger the concentricity of a need, the more effective is the effort of developing the element technologies required for the need, since more of these technologies will also be useful for other needs.

The concentricities of the needs in Table 3 are

$$Con(Y_1)=1.7,\ Con(Y_2)=2,\ Con(Y_3)=1,\ Con(Y_4)=1.7,\ Con(Y_5)=2.7$$

which are shown in Figure 3. The largest concentricity is assigned to need Y_5, which is seen to be very closely related to the other needs in Figure 3, and the smallest concentricity is assigned to the isolated need Y_3. Concentricity can also be computed for element technologies and their combinations. The concentricities are shown for the technologies and their combinations in Figure 3.

(b) Eccentricity [8] : Eccentricity $Ecc(Y_i)$ of need Y_i is defined by

$$Ecc(Y_i) = \frac{q - q'}{q'} \quad (3)$$

where q is the number of element technologies Y_i requires, and q' is the maximum of the number of element technologies Y_i shares with some other need.

Eccentricity is a measure that indicates how much a need is isolated from the other needs in scope with respect to sharing technologies. The larger the eccentricity of a need, the more is the need isolated from the other needs. In particular, a need that requires no common technology with other needs is assigned an eccentricity of infinity.

The eccentricities of the needs in Table 3 are

$$Ecc(Y_1)=2,\ Ecc(Y_2)=1.5,\ Ecc(Y_3)=\infty,\ Ecc(Y_4)=2,\ Ecc(Y_5)=0.5$$

The ascending order of these eccentricities corresponds to the descending order of the above concentricities for this example. In general, however, concentricities and eccentricities measure different structural aspects [7], and both measures are useful for strategy analysis.

(c) Utility indicator [10] : Utility indicators measure the utility of developing element technologies and their combinations based on an F-Hierarchy.

Utility indicators of technologies depend not only on the number of needs like corresponding concentricities, but also on the number of paths connecting the technologies and the needs on the F-Hierarchy. The indicators, therefore, are useful for considering strategic paths on the F-Hierarchy and for determining priority rankings of technologies for the needs that are selected as targets of technological developments.

We describe basic ideas of computing utility indicators on the example F-Hierarchy in Figure 3. We assume that the utility indicators of needs Y_i are specified by their concentricities, since the concentricities indicate an aspect of utility of the needs in terms of diffusion effects of developing element technologies to other needs. Starting from these indicators, we go down the F-Hierarchy to compute utility indicators of combinations of technologies and those of technologies. Since node $\circ\langle X_1\ X_3\rangle$ represents two out of the six technologies required for need Y_1, the node's contribution to Y_1 is evaluated by their ratio 2/6, and its utility with respect to Y_1 is (2/6) x (utility indicator of Y_1) = (2/6) x (5/3) = 5/9. Similarly, the utility of $\circ\langle X_1\ X_3\rangle$ with respect to Y_5 is

(2/3) x (utility indicator of Y_5) = (2/3) x (8/3) = 16/9. We assume additivity of these utilities, and determine that the utility indicator of node **o** $<X_1\ X_3>$ with respect to all the needs in scope is 5/9 + 16/9 = 7/3 = 2.3. The utility indicators of all the technologies and their combinations are computed in a similar manner, and are shown in Figure 3.

4 Needs analysis for fifth generation computer systems

We applied the method in Section 3 to the data in Table 2, and studied relational structures of the needs and technologies to analyze implications of the relations to the development of application technologies of FGCS and to their R & D strategies.

4.1 Structural relations on a Q-Hierarchy

Figure 5 illustrates the Q-Hierarchy of the needs in Table 1. Each rectangular node indicates a need and technologies required for the need. For example, node N1(23 25 29 36) of level 4 presents need N_1 that requires information technologies S_{23}, S_{25}, S_{29} and S_{36}. Rectangles D1 and D2 indicate connecting nodes, which are illustrated by **o** nodes in Figure 4.

The Q-Hierarchy, as compared with the corresponding F-Hierarchy, is particularly suited to grasp a holistic relational structure of the needs, because the former is a simplified model of the latter and has a tree structure.

We first describe findings obtained from the Q-Hierarchy regarding overall relations of the needs.

1) All the needs except N_{11} and N_{14} are related to each other, and form a single relevance tree.
2) Need N_{11} (new information industries) of level 7 and need N_{14} (health care industries) of level 1 are singular needs in terms of technological developments, since the technologies required for each of these needs are required for no other need.

We can visually analyze connectedness among the needs by observing the joining of paths from top to bottom.

3) The most closely related needs are needs N_{12}(medical electronics for medical offices) and N_{17}(medical service systems) of level 6, which are connected by node D1 of level 5. This implies that five out of the six technologies required for one of the needs are also necessary for the other need. These two needs were identified by breaking down different bottlenecks as Figure 1 illustrates (need N_{12} was obtained from bottleneck B_3 and N_{17} was obtained from B_4), but they turned out to be closely related social needs as their descriptions indicate, thereby requiring almost the same information technologies. Another relation of interest is that need N_{17} is related to need N_{16}(robots for dirty work) through node D1 (need N_{12} is related to N_{16} through N_{17} as we will see later in the corresponding F-Hierarchy). Indeed, N_{16} requires a subset (S_{34},S_{35},S_{36}) of the technologies (S_{22},S_{23},S_{26},S_{34},S_{35},S_{36}) required for N_{17}. The interpretation for this relation is that 'robots for dirty work' will utilize the same information technologies as those for robots that will be utilized for automating medical services (N_{17}). Thus, although these two social needs may be apparently different, necessary integrated technologies to meet the needs are constructed from the same element information technologies.

4) The other interesting group of needs with close relations is group A in Figure 5. Group A consists of four needs, N_1(decision support systems for planning and staff activities), N_2(office automation for routine office work), N_{13}(education for senior citizens) and N_{18}(CAI for personal education), and is connected by node D2 of level 3. This implies that for each of the needs in this group there is another need within the group that requires the same three technologies necessary for the former. As for the case of N_{16} and N_{17} in 3), the two distinctly different types of social needs, business needs (N_1 and N_2) and educational needs (N_{13} and N_{18}), are very closely related from the viewpoint of the information technologies they require.

5) The branches of the needs discussed in 3) and 4) are connected with ten other needs at level 2 of the Q-Hierarchy, and all of these in turn are connected with eleven other needs at level 1 to form the overall relevance tree. Interpretations of technology sharing relations of these needs are possible as in 3) and 4), although they are less closely connected.

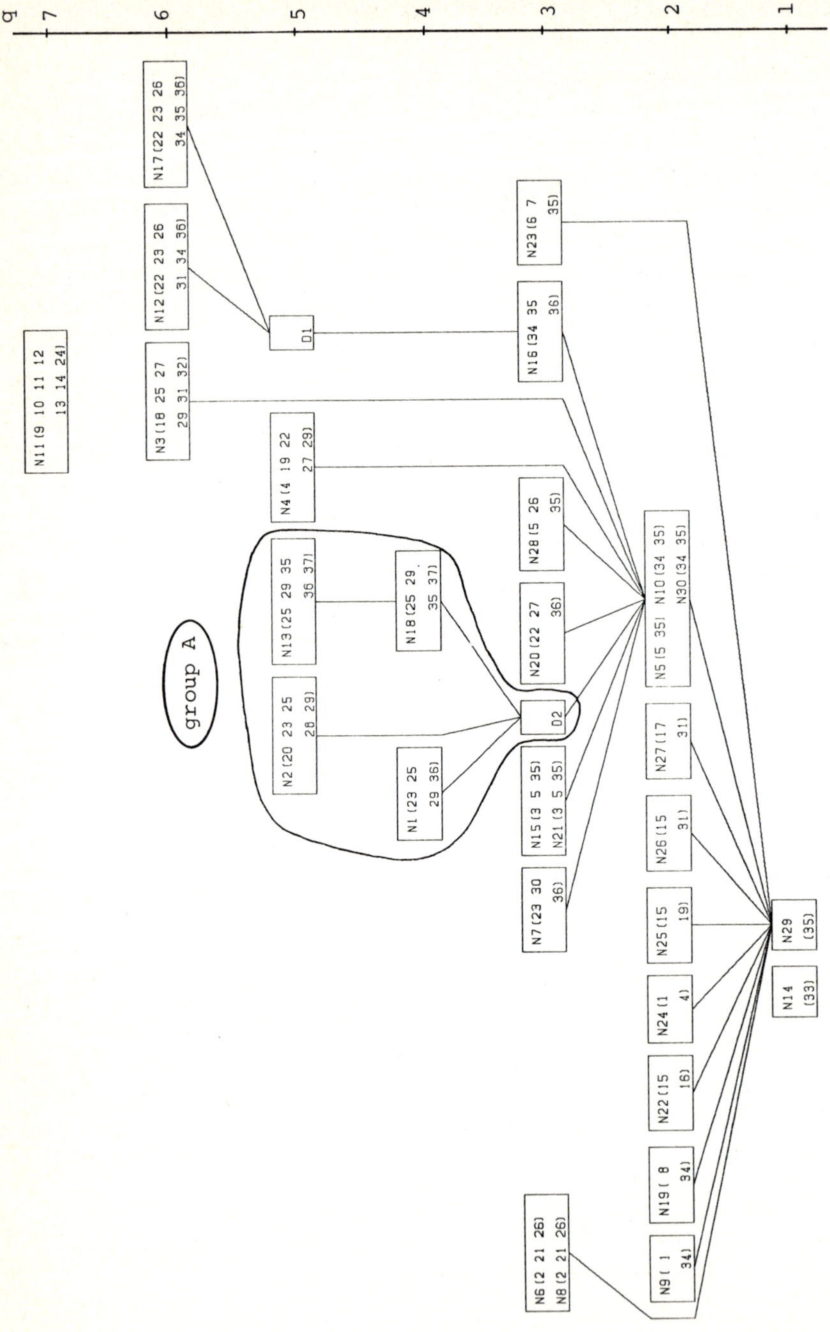

Figure 5 Q-Hierarchy of social needs

4.2 Structural relations on an F-Hierarchy

The F-Hierarchy corresponding to the Q-Hierarchy in Figure 5 is illustrated in Figure 6. The rectangular nodes Di(....) in Figure 6 represent o nodes of technologies and their combinations in Figure 3 that are not required for needs by themselves but are utilized for more than two needs together with additional technologies.

Since the F-Hierarchy explicitly illustrates the (element) information technologies necessary for the needs, it is useful for observing details of technology requirement relations of the needs. It is also useful for analyzing the importance of the element technologies and their combinations for the purpose of understanding how widely they will be utilized in meeting social needs. The eccentricities, concentricities and utility indicators in Table 4 and Figure 6 provide numerical measures for the analysis and assist us in finding promising application technologies of FGCS.

Table 4 Eccentricity, concentricity and utility indicator of nodes

	Ecc	Con	Utility indicator		Ecc	Con	Utility indicator		Ecc	Con	Utility indicator
N_1	0.33	5.75	5.75	N_{11}	∞	1	1	N_{21}	0.5	6	6
N_2	0.67	3.6	3.6	N_{12}	0.2	5.33	5.33	N_{22}	1	2	2
N_3	2	3.33	3.33	N_{13}	0.25	6.4	6.4	N_{23}	2	4.67	4.67
N_4	1.5	3.4	3.4	N_{14}	∞	1	1	N_{24}	1	2	2
N_5	0	8	20.67	N_{15}	0.5	6	6	N_{25}	1	2.5	2.5
N_6	2	3	3	N_{16}	0	8.67	12.01	N_{26}	1	3.5	3.5
N_7	0.5	4.33	4.33	N_{17}	0.2	6.67	6.67	N_{27}	1	2.5	2.5
N_8	2	3	3	N_{18}	0	6.25	11.37	N_{28}	0.5	7	7
N_9	1	4.5	4.5	N_{19}	1	4	4	N_{29}	0	12	41
N_{10}	0	9.5	17.51	N_{20}	0.5	4.67	4.67	N_{30}	0	9.5	17.51
D_1	-	5.33	6.47	D_9	-	4	4.45	D_{17}	-	2	3.25
D_2	-	5.5	16.54	D_{10}	-	3	3.48	D_{18}	-	7	12.01
D_3	-	6	9.51	D_{11}	-	4	5.8	D_{19}	-	4.8	4.8
D_4	-	6	9.78	D_{12}	-	3.5	4.47	D_{20}	-	5.5	7.11
D_5	-	6	8.15	D_{13}	-	2	1.68	D_{21}	-	9.5	10.57
D_6	-	7	22.47	D_{14}	-	2	1.93	D_{22}	-	8.5	6.89
D_7	-	5	7.05	D_{15}	-	5	7.45	D_{23}	-	3	4
D_8	-	4.5	2.47	D_{16}	-	7	27.78				

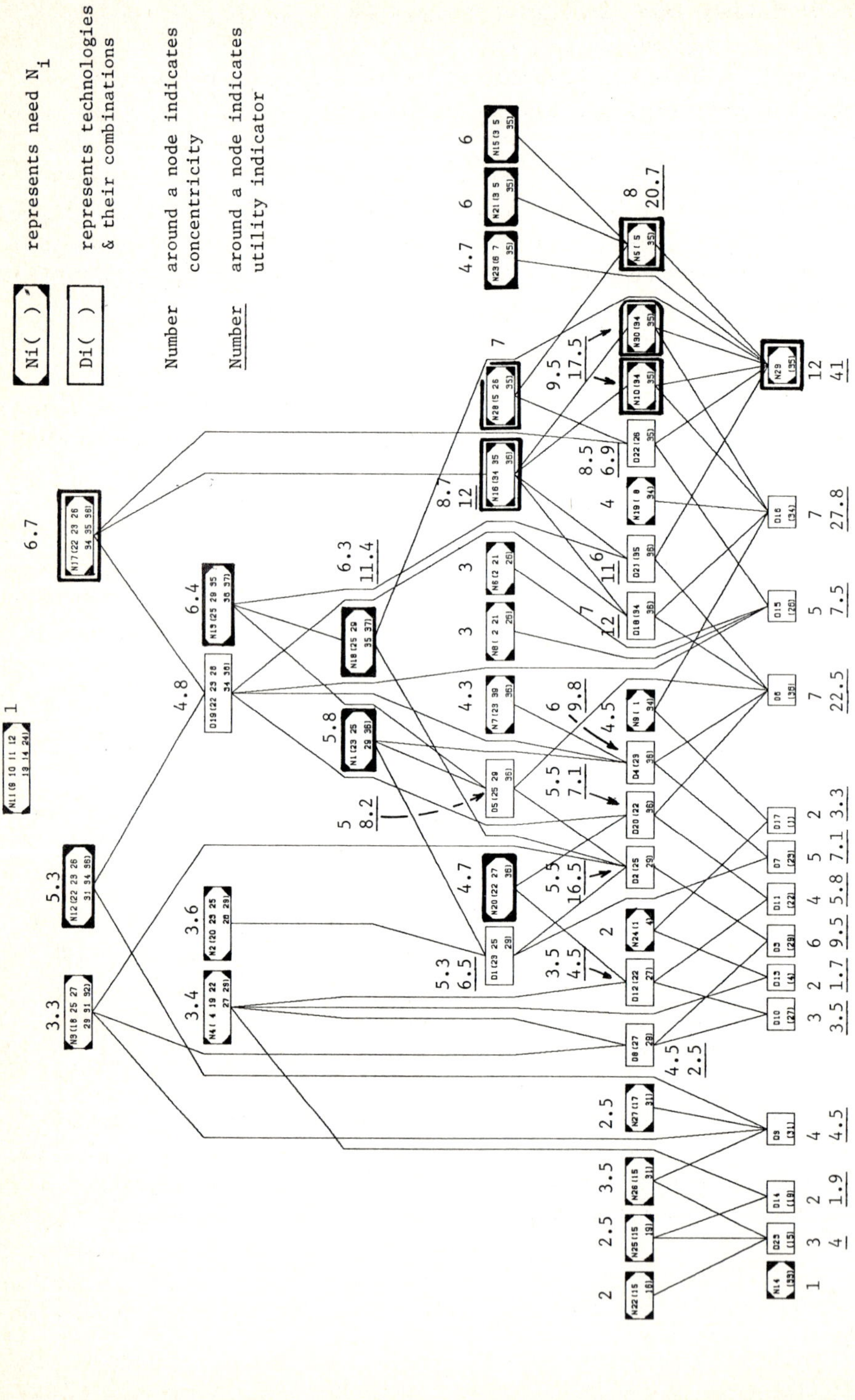

Figure 6 F-Hierarchy of social needs

We first investigate details of the relations observed on the Q-Hierarchy in the previous subsection.

1) The five technologies shared by the closely related needs N_{12} (medical electronics for medical offices) and N_{17}(medical service systems) are represented by D19(22 23 26 34 36) ; that is, information technologies for S_{22}(protection of data security), S_{23}(associative retrieval), S_{26}(picture and image processing), S_{34}(sensor bases) and S_{36}(databases) are required for both needs. These five technologies will be representative information technologies required for future medical needs.

2) The branch of group A in Figure 5, which illustrates close relationship of needs N_1(decision support systems for planning and staff activities), N_2(office automation for routine office work), N_{13}(education for senior citizens) and N_{18}(CAI for personal education), is disaggregated in Figure 6 and illustrates that N_1 plays a key role in this group. Need N_1 is a key, since on the one hand it shares with need N_2 technologies S_{23}(associative retrieval), S_{25}(picture processing) and S_{29}(Japanese language processing) represented by D1(23 25 29), and on the other hand it shares with need N_{13} technologies S_{25}, S_{29}, and S_{36}(databases) represented by D5(25 29 36). Since need N_{18} can be met by a subset $\{S_{25}, S_{29}, S_{35}, S_{37}\}$ of technologies required for N_{13}, need N_1 is located in the central position of a chain of technology sharing relation N_2-N_1-N_{13}(N_{18}) ; i.e., information technologies for decision support systems will also be quite useful for educational needs as well as for office automation.

The F-Hierarchy also illustrates element technologies and their combinations that are promising in meeting social needs. The nodes of lower levels in Figure 6 that are connected to higher level nodes with more edges are promising, since they will be utilized in more integrated technologies for the needs. The concentricities and utility indicators shown in Figure 6 and Table 4 are useful measures to help us find promising technologies on the F-Hierarchy ; the greater the concentricity/utility indicator of a node, the more promising is (are) technology (technologies) represented by the node.

3) The most promising element technologies among those represented by nodes of level 1 are, S_{35}(built-in microcomputers) of node N_{29}, S_{34}(sensor bases) of node D_{16} and S_{36}(databases) of node D_6, which are nodes of the top three utility indicators. Element technologies S_{29}(Japanese language processing) of node D_9, S_{26}(picture and image

processing) of node D_{15}, and S_{23}(associative retrieval) of node D_7 are also assigned large utility indicators and hence are promising.

4) Combinations of two element technologies are of particular interest in this analysis. Figure 6 and Table 4 indicate that the following combinations are promising ; S_5 and S_{35} of node N_5, S_{34} and S_{35} of nodes N_{10} and N_{30}, S_{25} and S_{29} of node D_2, S_{34} and S_{36} of node D_{18}, S_{35} and S_{36} of node D_{21}, S_{23} and S_{36} of node D_4, and S_{26} and S_{35} of node D_{22}. These combinations are generally considered as those information technologies which will play important roles for constructing computers of easy access and of intelligent functions for human users, and hence are promising application fields of knowledge information processing systems.

5) For combinations of three element technologies we can similarly observe that the developments for (S_{34},S_{35},S_{36}) and (S_5,S_{26},S_{35}) are effective since these technologies are required for various needs other than need N_{16} for the former and need N_{28} for the latter.

The element technologies and their combinations discussed in 3), 4) and 5) are important information technologies to meet anticipated social needs of Japan in the 1990s. Therefore, they will become important application technologies of FGCS.

4.3 Strategy analysis

The F-Hierarchy in Figure 6 provides us with a solid model for considering strategies for developing application technologies, since it illustrates the paths by edges connecting Di(....) nodes of element technologies and their combinations to need nodes of upper levels for which they are utilized together with additional element technologies.

We can observe that the needs of the top seven concentricities (the top quarter of all the needs) in Table 4 are located at the right hand side of Figure 6 from level 1 to level 6 (these nodes are shown with double-lined rectangles). These seven needs constitute a closely related group of needs and may be considered as a strategic core of application fields of FGCS, because Figure 6 illustrates cross support relations of the needs through the technologies they require and the developments of element technologies for these needs will be quite effective as their concentricities indicate.

The needs with the 8th to 15th concentricities (the second quarter of all the needs) are illustrated in Figure 6 with rectangles of bold lines. This group is more scattered in the F-Hierarchy than the core group, but we can see that it is scattered around the core group.

Thus the relational model represented by the F-Hierarchy advises the following effective strategy for developing integrated technologies to meet the needs from the combinatorial view point of information technologies :

1) The needs in the core group are top priorities for developing integrated technologies as application fields of FGCS.
2) Those of the second quarter are secondary, and the development of technologies for those needs should be assessed in relation to the core group.
3) The rest of the needs are then considered in relation to the above needs.

A method has been devised to determine priority rankings of the element technologies for developing integrated technologies which correspond to targeted needs [10]. This method determines the rankings on a hierarchical graph generated by adding all the element technologies to an F-Hierarchy. Its procedure is to trace edges on the graph from nodes of level 1 representing element technologies up to the nodes representing the target nodes, where utility indicators are utilized as a measure to determine directions of the tracing. Let us consider the case where N_{17} is specified as the target, which is the highest level need among the needs in the core group. The priority ranking of the element technologies needed to develop the integrated technology required for need N_{17} can be determined by the method, and is specified as

$$S_{35} > S_{34} > S_{36} > S_{26} > S_{23} > S_{22} ,$$

where $S_i > S_k$ denotes that technology S_i is given a higher priority for development than technology S_k. Thus, S_{35} should be given the highest priority and S_{22} the lowest priority. Priority rankings of technologies required for more than one target need can also be determined by this method.

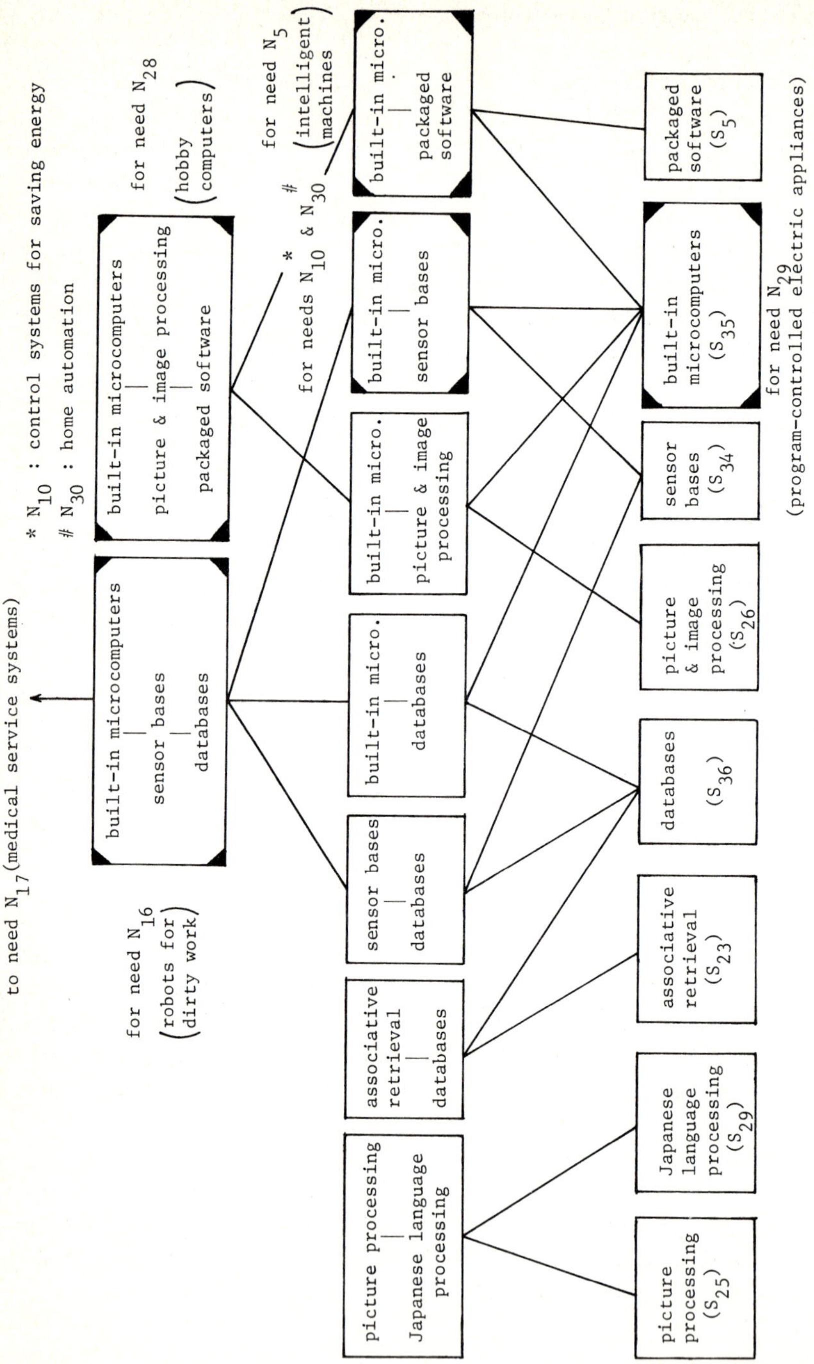

Figure 7 Important application technologies and their relations

4.4 Important application technologies to meet social needs

We can identify important information technologies by summarizing the observations in Subsections 4.2 and 4.3. In particular, the promising element technologies and their combinations observed in 3)-5) of Subsection 4.2 and the core of application fields of FGCS observed in Subsection 4.3 can be integrated to yield Figure 7, which extracts important element technologies and their combinations out of Figure 6 and illustrates the technologies explicitly. Figure 7, which also illustrates the core of application fields, and the observations in Subsections 4.2 and 4.3 enable us to describe the following scenario for the development of key application technologies to meet the anticipated social needs in the 1990s.

Technologies for built-in microcomputers will be the most important for meeting various aspects of social requirements. To start with, microcomputers will be built in all kinds of electric appliances to enable their control according to flexible programs. Integration of other promising information technologies with built-in microcomputers will be quite useful for many application fields ; integration with packaged software for constructing intelligent machines, and integration with sensor bases for constructing control systems for saving energy as well as for various aspects of home automation. Built-in microcomputers and packaged software will also be utilized for hobby computers with an addition of technologies for picture and image processing.

Database technologies will also be quite important. Database technologies alone will not be used for application technologies, but a great deal of application fields are in the perspective by integrating them with other information technologies ; associative retrieval, sensor bases and built-in microcomputers. As an example, technologies for databases, sensor bases and built-in microcomputers will be major element technologies for constructing intelligent robots that will replace human labors for dirty work. These robots will alleviate labor shortages due to structural changes in Japanese society. The three element technologies will also be utilized for constructing effective medical service systems, which will be needed for the nation in view of the rapid increase of the elderly.

Another group of important technologies will be those for processing

Japanese language and pictures. These technologies will be required for office automation and computer aided instructions, which will help improve productivity in the areas of office work, service sectors and education.

The application technologies identified above will be important information technologies to help alleviate the anticipated social needs of Japan in the 1990s. Also these technologies will play an important role in developing knowledge-intensive technologies that will contribute to the development of a technology-based nation.

5 Concluding Remarks

We have discussed the results of a needs-oriented analysis for research and development of fifth generation computer systems(FGCS). We described briefly Japanese R & D efforts for the computer systems and a needs analysis that identified necessary information technologies required for anticipated social needs. We presented a brief description of the method of analysis originated for this study, which consists of Visual Q-Analysis and strategy analysis. We applied the method to the relational data of the above needs analysis, and obtained hierarchical relational structures of the needs for FGCS in the form of a Q-Hierarchy and an F-Hierarchy. The hierarchies were useful tools for visually understanding relations of application fields of FGCS, and we have identified a strategic core of the needs and important application technologies of FGCS. Furthermore, we have presented a scenario for the development of the key application technologies.

Identification of needs for FGCS and relevant technologies is very important to obtain meaningful results of the needs-oriented analysis, so that the analysis is useful for considering R & D plans for application technologies of FGCS and related information technologies. For this purpose it is necessary to update and refine the needs analysis summarized in Tables 1 and 2 as social environments change and the R & D plans are refined. Indeed, a needs analysis and R & D plans should be updated in an evolutionary manner, since they impact each other as the state of society and the state of the art of technologies evolve in time. Another direction to refine the analysis for R & D of FGCS is to analyze explicit relations among application oriented information technologies like those in Table 1 and basic computer techniques for FGCS specified in the R & D plans of ICOT [4].

The method for this study will prove to be useful in an early stage of planning technological developments, as we have shown for the fifth generation computer systems. Available data for evaluating technologies of the planning are limited at this stage. For this stage Visual Q-Analysis will be useful in presenting a model of the scope of planning. Strategy analysis will also be useful in considering strategies for technological developments on the model, since the method is based on limited relational data among needs and technologies. Various other aspects like time for developments need to be taken into consideration for the planning ; therefore, this method should be utilized together with other methods that account for such aspects. Further research is envisaged to integrate these methods with other methods [e.g.11] that enable more detailed analysis when more information such as costs, time, utility and probabilities are available regarding technological developments. Such an integration of methods will be useful as methodology to support project management of research and development of innovative technologies such as those for the fifth generation computer systems.

REFERENCES

[1] Moto-oka, T. ed., Fifth Generation Computer Systems : Proceedings of the International Conference on Fifth Generation Computer Systems, Tokyo, October 19-22, 1981 (North-Holland, Amsterdam, 1982).

[2] Ministry of International Trade and Industry, The Vision of MITI Policies in 1980s - Summary -, News from MITI NR-226(80-7) (March 17, 1980).

[3] Japan Information Processing Development Center, Investigation Report on the Social Environments of the Fifth Generation Computers, Report 54-R013 (March 1980) (in Japanese).

[4] Institute for New Generation Computer Technology, Outline of Research and Development Plans for Fifth Generation Computer Systems (May 1982).

[5] Institute for New Generation Computer Technology, Institute for New Generation Computer Technology - Introduction -,(June 1982).

[6] Sugiyama, K., M.Toda and S.Tagawa, Visual Q-Analysis and Its Application to a Structural Study of Technology Innovations, Research Report No.19, International Institute for Advanced Study of Social Information Science (July 1981).

[7] Sugiyama, K. and M.Toda, Visual Q-Analysis : A Case Study of Future Computer Systems Developments, Collaborative Paper CP-82-51, International Institute for Applied Systems Analysis (August 1982).
[8] Atkin, R.H., Combinatorial Connectivities in Social Systems (Birkhäuser, Basel, 1977).
[9] Sugiyama, K., S.Tagawa and M.Toda, Methods for Visual Understanding of Hierarchical System Structures, IEEE Trans. Systems, Man and Cybernetics, SMC-11 (1981) 109-125.
[10] Toda, M., A Method based on Structural Analysis for Evaluating Strategies of Technological Developments, Research Report No.31, International Institute for Advanced Study of Social Information Science (September 1982).
[11] Cetron, M.J., J. Martino and L. Roepcke, The selection of R & D Program Content - Survey of Quantitative Mehtods, IEEE Trans. Engineering Management, EM-14 (1967) 4-13.

R & D Management Systems in Japanese Industry
H. Eto and K. Matsui (editors)
© Elsevier Science Publishers B.V. (North-Holland), 1984

Chapter II TOWARD A POLICY GUIDANCE SYSTEM FOR COMPLEX INNOVATION

YUSAKU SHIBATA

Planning Group
Systems Development Laboratory
Hitachi, Ltd.
Ohzenji, Asao-ku, Kawasaki, Japan

By analyzing the obstacles in complex innovations, it has been made clear that the most critical deficiency in management is the shortage of policy-relevant information for action-oriented managers and professionals. A new policy-relevant communication system is proposed here and named the "policy guidance system (PGS)". Experiences of one company were analyzed and it was found that various policy guidance functions are already emerging.

1 INTRODUCTION

"Uncertainty" and "complexity" are buzz words in innovation management. In order to make the innovations successful , companies must operate under focused strategies but yet have flexible management systems. The key factor in this effort is the structural change of institutions based on a new paradigm or way of thinking. It is said that the difficulties and risks in institutional change will be especially formidable for Japanese companies.

> "... the ability to innovate and explore the uncharted is precisely what separates the leader from the industry follower. The Japanese system has shown great strength in coming up from behind. But the kind of organizational skills that were needed

for catching up are probably not appropriate for being at the cutting edge. In time, Japanese society and industrial organizations may transform themselves in order to fulfill these new tasks. Until then, non-Japanese competitors have real opportunities to hold their own by exploiting Japanese institutional rigidites." [1]

Peter F. Drucker, however, has a completely different opinion:

"What is fact is that the secret behind Japan's economic achievement is not a mysterious Japan, Inc., a creation that belongs, if anywhere, in some Hollywood grade B movie. Far more likely, it is that Japan —at present alone among the major industrial nations— has addressed herself to defining the rules for a complex, pluralist society of large organizations in a world of rapid change and increasing interdependence." [2]

However, these seemingly contradictory opinions indicate an important common point, that is, many industrial organizations, regardless of whether they are Japanese or non-Japanese, are adapting themselves to a world of rapid change and increasing interdependence.[3] This organizational adaptation itself is another innovation which is probably far more difficult and important than technology-oriented innovations. Complex innovations or the exploration of uncharted territory will come up against many superficial and underlying obstacles. This makes a new guidance system not only desirable but also indispensable.

This article will describe the features of complex innovation, the obstacles which necessitated this new guidance system, a conceptual framework of PGS and an experimental approach to the implementation of the PGS concept in Japan.

2 TYPES OF TECHNOLOGICAL INNOVATIONS [4]

Technological innovations are not single simplistic events but are composed of many complex and fundamentally different processes.

Therefore, it is necessary at first to consider the different types of technological innovations and define the field of discussion. There appear to be at least four basic categories that are useful for this purpose:

(1) Marginal shifts that improve existing processes and products, such as the automated operation of thermal power plant by using a computer.

(2) State of the art advances that entail the straightforword replacement of one type of equipment with another, such as a gas-turbine power plant for a boiler/steam-turbine power plant.

(3) State of the art advances that entail fundamental displacement or substitution of one item by another, such as nuclear fuel over fossile fuel.

(4) New areas of innovational activity and development, such as solar power satellites and soft energy paths.

In addition to these four categories, there is another context which must be considered with respect to the complexity of the innovation process. The following three categories seem to be appropriate for discussing this context:

(1) Simple innovations that are achieved largely within existing paradigms and institutional mechanisms, and are determined predominantly on technical and economic grounds.

(2) Complex innovations that are achieved only by injecting them into new economic, social, or technical areas. These innovations require new forms of cooperation or market aggregations. In these innovations, technical and economic factors are important but may be less of a determinant than are other obstacles.

(3) Super-complex innovations are achievable only through major societal reforms which might require a coordinative

> strategy of public and private institutions. Accordingly, in the achievement of these types of innovations, while technical and economic considerations are formidable and highly complex, they may be relatively less difficult to overcome and less of a determinant of whether or not the innovation is realizable than the social and political dynamics of the process.

In this article, the discussion will be focused on the more complex process of innovations in new areas. New energy technology developments appear to belong in these categories.

3 OBSTACLES TO COMPLEX INNOVATIONS

Managers and professionals, who are engaged in such complex innovations as new energy developments, must overcome many obstacles, both technical and non-technical. Their actions will possibly become more rational and effective if relevant information and guidelines are timely offered to them.

Since 1980, the Engineering Advancement Association of Japan (ENAA) has been studying the feasibility of an energy information service system which will assist managers and professionals in exploiting opportunities and addressing issues related to energy developments. As a part of this study, major obstacles to energy developments in Japan have been gathered and analyzed by a committee, members of which are Prof. S. Hirayama of Ikutoku Institute of Technology, Mr. K. Konno of New Energy Development Organiaztion, Mr. K. Tachibana of Tokyo Electric Power Co., Mr. T. Suzuki of International Energy Forum, Mr. Y. Kitaya of Japan Management Association, Mr. H. Imoto of Imoto Engineering Co., Mr. O. Wada of Engineering Advancement Association of Japan and the author.

Twenty-four major obstacles were raised by the committee members and categorized into eight groups as described below. Eight of these obstacles which appeared to be the most relevant to the concerns of the committee were used as the titles for the following descriptions of obstacles. Further, causal relations between these eight obstacles were analyzed by the committee utilizing a simplified Interpretive Structural Modeling (ISM) method [5] and shown in Figure 1.

A. UNCERTAIN RESOURCE SUPPLY AND MARKET DEMAND: Japan is now developing new energy technologies under the leadership of the national government. Although it is clear that limited capital and human resources should be concentrated on the most promising projects, the actual selection is extremely difficult. For example, there are at least five promising coal gasification technologies for power generation, all of which were demonstrated to be chemically and thermodynamically feasible, but are still uncertain from the standpoints of economics, adaptability to various kinds of coals, possibility of increasing the scale, and impact on the environment.

Moreover, forecasting of the market, which includes, the growth of economic activity and energy demand, the supplied quantity and price of oil and natural gas, and the siting of nuclear power plants, is an indispensable premise in planning, and is even more difficult. Because such a climate of uncertainty makes investment decisions of private companies difficult, it is said that the government agencies must consider providing funds, both R&D support and market subsidies in order to realize benefits for society at large.

B. DIVERSITY OF VALUES AND NORMS: Values and norms are important factors which are closely related but not identical with the market. On the one hand, viewed from the perspective of economic growth oriented people, the energy problem is essentially a problem of developing new supplies to meet the expanding energy needs of a dynamic society and the entailing irreducible environmental impacts are the price of maintaining a strong economy. On the other hand, viewed from the perspective of environmentalists, the energy problem is but one manifestation of a very fundamental crisis involving

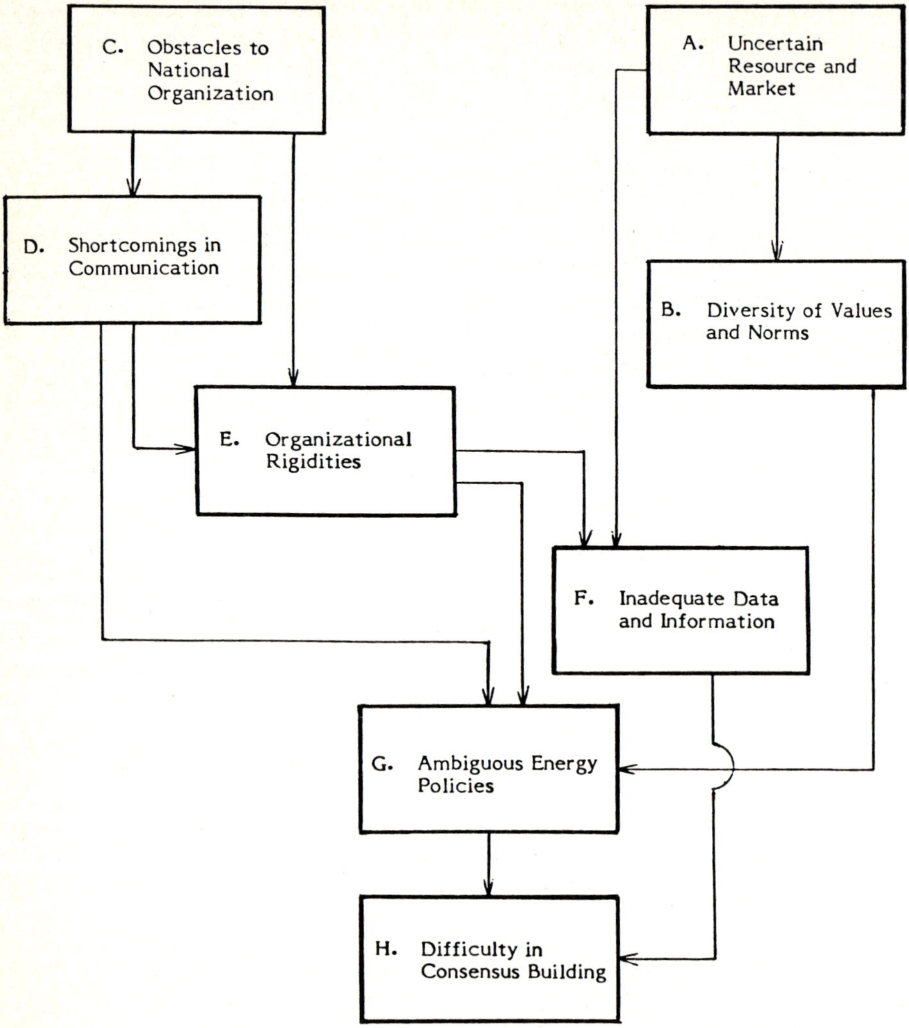

Figure 1
Relational Structure of Obstacles to Energy Developments

nothing less than the future survival of industrialized society. The outrage due to environmental degradation demonstrates the need for a new social order which integrates these diverse values.

C. OBSTACLES TO NATIONAL ORGANIZATION: Complex innovations require a high degree of cooperation between public agencies and private companies that generally does not exist unless a catalytic mechanism intervenes in the process. Although the New Energy Development Organization (NEDO) and the New Energy Foundation (NEF) have been established as new national organizations for the development of new energy technologies, the working of these institutions depends on the related organizations' desire for achieving their goals. It also depends on their mutual trust and ability to agree on the aims and methods of concerted action.

D. SHORTCOMINGS IN COMMUNICATION: Inter- and intra-organizational cooperation depend on a well organized transmission of information which meets the needs of members. In order to satisfy such needs, the communication system has to be constantly evolving to meet new requirements and aspirations such as the energy problem and new values. However, the complexity of technical jargon and the ambiguity of Japanese conversational style together with the pressures of the mass media obstruct smooth communications.

E. ORGANIZATIONAL RIGIDITIES: Every organization tends to become rigid as a result of specialization and bureaucratization. It is increasingly felt that present organizational mechanisms are not effectively adapting to the needs of complex innovations. The underlying problem is the difficulty of restructuring the traditional organization while keeping the efficiency of the present operation.

F. INADEQUATE DATA AND INFORMATION: Although the three fundamental features of the Japanese energy policy ——

securing a stable supply of petroleum, developing alternative energy resources and promoting energy conservation — have already been decided on, the implementation strategy of this policy has not yet been established. One of the major obstacles which is hindering the establishment of an implementation strategy is the shortage of effective information. If we regard information as synonymous with quantitative data, such as production, import and consumption of energy data, it would be difficult to conceive of a basic shortage of information. However, the meaning of shortage is the lack of effective information relative to needs and capabilities. It is necessary to distinguish between the concept of information and the concept of data. Information is the organization and reordering of data, such that these processes interrelate the data base with political, economical, and social realities. Therefore, with all the energy related data and information that are available, there are still widespread complaints calling for effective information by which to formulate energy strategies.

G. AMBIGUOUS ENERGY POLICIES: Action-oriented managers of energy-related projects, who operate under the complexity of all the obstacles to innovation, need guidelines for interrelating their action plans within the scope of total policy considerations. However, a peculiarity of Japanese management is that it takes a longer time for starategic decision-making because a lot of time is spent considering the decision among many managers and colleagues without having explicit top-down policy guidelines. Therefore, the essential need will be for an improvement in the Japanese decision-making process and/or the creation of a policy guidance system.

H. DIFFICULTY IN CONSENSUS BUILDING: The solution to the energy problem will come about as a result of a combination of the strong motivation and social behavior of the Japanese people. There are many mechanisms such as the government, education system and mass media that either stimulate or limit the understanding and motivation of people. It is neccessary but very difficult for these social mechanisms to adapt themselves to the needs of the energy crisis and diversified social values. As a result, the siting of every power plant meets with resistance from concerned citizens. At the same time, the promotion of energy conservation is usually neglected by indifferent people.

4 SELECTED INITIATIVES FOR POLICY GUIDANCE

In the 1960s, the divisionalization of company organizations was introduced in order to decentralize product oriented operations. When strategic planning , which is oriented to the market and covers more than one division became necessary in the 1970s , companies tried to recentralize the planning process usually at a higher level than that of divisions. In addition to these operation and strategy problems, the complex innovations of the 1980s raise new policy problems such as the energy problem noted previously. In the same way as companies relied on new concepts such as SBU (strategic business unit) to provide a strategy planning framework, they must develop another concept to provide policy guidance for managers and professionals who are engaged in these complex innovations. From the preceding section, it is clear that the policy guidance system (PGS) must play a variety of roles with respect to the obstacles facing innovations. However, because the obstacles are highly interdependent, the policy guidance framework cannot be such that there is only one function for one obstacle. Rather, the framework that should emerge is one of a balance, or a mix of functions to deal with multiple obstacles and interrelationships at the various levels of management as shown in Figure 2. The following description of the five major functions appears to be useful in defining the conceptual framework of PGS:

(1) FUTURE FORECASTING AND RISK ANALYSIS: Based on research of future social and technological environments, PGS furnishes alternative patterns or scenarios of the

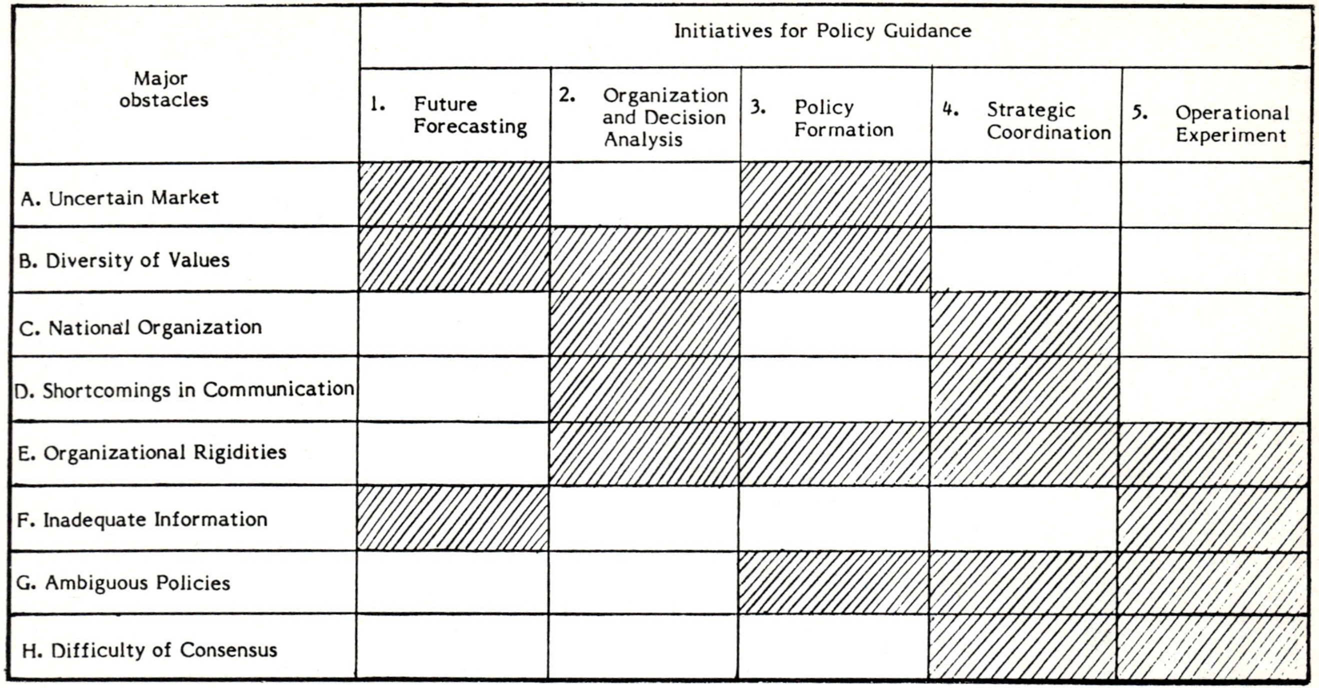

Figure 2

Illustration of Interactions between Guidance Roles and Obstacles

possible future.

(2) ORGANIZATION AND DECISION ANALYSIS: PGS catalyzes a proactive restructuring of an organization in order to adapt it to a possible future and normative policies.

(3) POLICY FORMATION: Using a future scenario and the restructured organization as background, PGS helps top managers establish normative objectives and policies.

(4) STRATEGIC COORDINATION: Based on the experience of policy formation, PGS helps middle managers select goals and plan strategies. Furthur, PGS facilitates the resolution of conflicts between policies, strategies and tactics.

(5) OPERATIONAL EXPERIMENT: The above-mentioned experiences of higher level planning are utilized by PGS to help operating managers solve daily problems. In turn, this experience will be fed back and used in higher level planning .

4.1 Future Forecasting and Risk Analysis

The proces of innovation occurs in the present, but its real objectives are future benefits. Therefore, the decision-maker, who must make decisions now expecting returns in the future, typically asks himself three questions: What is likely to happen without specific intervention (forecasting)? What should be our aims (policy)? How should the desired future state be brought about (tactics)? The first question demands a forecasting and is addressed to market and technical researchers. The second, is essentially normative in content and is addressed to top management. The third, which is concerned with means, is addressed to professionals.

If the situation is very simple like that of Robinson Crusoe, the answers to these questions will also be very simple: Nothing will happen without his intervention (short-term forecasting). He should and can decide his goals for the future (policy). He must choose his actions and do his best (tactics). Although this example seems too simple to have any practical meaning, it is still the essence of all personal decision-making. After gathering data from various sources and transforming them into a pattern, he selects his action intuitively based on visceral judgement. Thus, the forecasting, the policy making and the selection of tactics are inseparably intermingled in personal decision-making.

However, a complex innovation requires a long lead time, the cooperation of many people and large amount of money, which in turn, demand a clearly established strategy, a consensus among related people and a cooperative organizational structure. As a result, the decision-maker will lose sight of the patternized understanding of the situation and will have no solid ground on which to base his decisions. This is because the related data become too enormous and diverse to be handled by one person. Looking for ways to reduce this confusion, he calls on experts for forecasts in the hope that this will simplify the problem so that he can again use his intuitive reasoning.

We must be cautious in responding to this request, because this seemingly simple request for a forecast includes two independent problems: quantitative (forecasting) and qualitative (patternizing). Although there are already many econometric and energy models and forecasted results, past experience shows that forecasting, at least in the economic and energy areas, is still more of an art than a science. In fact, the track record of subjective forecasts has generally been more accurate than most of the available mathematical models. Therefore, these mathematical and subjective methods cannot come up with a single correct pattern of the predetermined future. However, they can be used to reduce the number of possible futures to some manageable number. Each of these possible futures reflects a coherent combination of quantitative and qualitative variables.

Thus, by using such a multiple but manageable number of possible future patterns, the decision-maker again can select his action intuitively based on visceral judgement. The problem still remaining, is how to build the consensus and secure the cooperation of the people. Because many people in the organization participate in the planning and implementation process of a complex innovation, how to integrate their personal opinions and establish a strategic plan is a key issue, and will be discussed next.

4.2 Organization and Decision Analysis

The obstacles to be solved next, are those caused by organizational rigidity and are summarized as follows:

(1) Obstructions to the formation of policy.

(2) Obstructions to the planning of strategy.

(3) Obstructions to the management of strategy.

These obstacles are caused by the following bureaucratic characteristics of organizations:

(1) Profits resulting from innovations are more difficult to evaluate than short-term profits.

(2) Intra-company competition between profit centers obstructs cooperation.

(3) The complexity of technical jargon makes communication and mutual understanding difficult between colleagues.

In order to overcome these obstacles, PGS can take advantage of the abundant knowledge of applied behavioral sciences. [6] In solving these problems as members of a task force team, the staff of PGS will function as linking pins between various parts of the organization. As a result, the structure will adapt to the new environment and strategy. Thus, from the viewpoint of behavioral sciences, the role of PGS is a catalyst for organization development.

4.3 Policy Formation

As already discussed in the previous chapter, action-oriented managers of complex innovation projects need guidelines for interrelating their decisions within the framework of total policy considerations. Therefore, the main activity in policy formation is the definition of normative objectives. These are defined in light of the value analysis of the forecasted consequences of any action being considered. Basically, the policy should reveal what ought to be done, under a given situation and in view of the accepted future scenario. In other words, policies can be considered the guidelines, and PGS, as the guideline of these guidelines.

Figure 1 implies that the direct causes of the "ambiguous energy policy" are the "diversity of values and norms", the "shortcomings in communication" and the "organizational rigidities". In turn, the "ambiguous energy policy" and "inadequate data and information" are causing "difficulties in consensus building". Therefore, PGS will help top managers establish normative objectives and policies by which action-oriented managers can overcome the obstacles of value diversity, communications shortcomings and organizational rigidity and interrelate their decisions within the total system. This description reminds us that such a function is already embedded in society, at least in Japan, as one American observer reports:

> "... Japanese manufacturers also realize that decisions at the level of operations can, if handled in a wise and consistent manner, have a useful cumulative effect at the level of strategy. Experience taught the Japanese the value of placing even short-term manufacturing decisions at the service of long-term strategy — a lesson that American companies have learned only imperfectly." [7]

The remaining challenge to Japanese managers is to transfer this lesson from the factory floor to the executive office where decisions are made at the level of strategies for technological innovations. This task will be especially enormous for the Japanese because they are not accustomed to think and discuss openly abstract things like future scenarios, and serious problems such as organizational obstacles and vulnerabilities. PGS will assist managers by offering information and tools to handle these problems.

4.4 Strategic Coordination

The bulk of the coordination efforts in the area of technological innovations is done at the strategy level where diverse alternative plans are coordinated in accordance with the policy guidelines. Although strategic coordination is very difficult because the risks are high, uncertainties great and potential returns so distant, corporate management cannot avoid making choices now. In strategic coordination, the basic questions are very simple:

(1) Which technologies are promising investment opportunities?

(2) How much should be invested in each of these technologies?

To answer these questions, however, is extremely complex and difficult. For example, in order to select a promising coal gasification technology for power generation from among multiple choices, strategic planners must study the present state and future potential of the economics, adaptability to various coals, possibility of increasing the scale and the impact on the environment of all of these technologies. Further, they must compare these features with those of competing technologies such as pulverized coal combustion, fuel cell technology, magnetohydrodynamics and nuclear technology. As for optimum investment, they must compare the proposed investments and expected profits based on market forecasts.

As technological innovations become more and more complex, covering multiple business units, the strategic coordination begins to impose a heavy sometimes unacceptable burden on top management. However, because these coordination activities require a broad range of vision and can significantly affect the company's long-term competitive strength, top managers cannot delegate the responsibility of coordination to their line subordinates whose views are usually narrower and more short-sighted. Consequently, top management must devise new mechanisms for dealing with strategic coordination based on corporate policy. While maintaining the comprehensiveness and thoroughness of strategic business planning, strategic coordination of complex innovations should avoid the bureaucratic formalities of strategic business planning. This is because they tend to degenerate into meaningless exercises, punctuated by ritualistic meetings and thick reports that neither inform top managers nor help operating people get their jobs done.

4.5 Operational Experiment of Strategic Coordination

In Japan, many people at the operation level try to solve operational problems considering their effect on the broad strategy. As the innovation process becomes more complex, this approach will become more difficult and operating managers will, thus, more often seek the advice of PGS. Further, because the problems at the operation level are generally simpler and easier than those at the strategy level, it is more prudent for PGS to initially be used in tackling mainly operational problems.

Strategic coordination at the operation level is useful not only as experimental operations in the developmental stage, but also as a source of necessary information to be fed back as inputs to the strategy level at a later stage.

5 TOWARD A STRATEGY FOR POLICY GUIDANCE

As has already been made clear, the unique problems in complex innovations are caused by the co-existence of complexity and innovation. However, the management of complexity and innovation itself is not new. For example, multi-national corporations are managing complexity fairly well. On the other hand, small venture companies are examples — successful and unsuccessful — of the management of innovative processes. As for the problems caused by the co-existence of these two very different activities, many engineering companies are organized along the lines of matrix structures, which facilitate the management of two different activities — project coordination and special functions. Therefore, the uniqueness and basic issues of PGS consist in how to resolve three contradictory problems: complexity, innovation and rapid change in the environment.

Now, the problem to be faced in addressing how PGS might establish itself in its relationship with other corporate activities is whether it will attempt to deal with the full complexity of the issue or, in the interests of practicality, will seek only modest incremental and basically low-profiled initiatives. The basic question is how PGS or the corporation as a whole will seek ways to master the complexity of fundamental change rather than to be mastered by it. That is one of the pervasive challenges to the contemporary world for which there is no clear-cut solution and will require of us a great deal of courage and creativity .

Consequently, the strategy that is emerging in Japan is one of a modest, but balanced mix of catalytic roles which simultaneously deals with multiple obstacles and their interrelationships . The organizational center of PGS is essentially a PGS task force which is composed of appropriate members selected from PGS research groups and PGS user groups or operating divisions. The PGS task force is an exploratory center which will serve as a neutral meeting ground for various managers and professinals from the operating divisions who must solve strategic problems while interrelating their decisions based on policy considerations.

6 EXAMPLE IN A JAPANESE ENTERPRISE [8]

6.1 Background

Hitachi, Ltd. is one of the most diversified technology-based companies in the world. Recently, it is endeavoring to respond to new trends toward complexity in the market by integrating its accumulated and new technologies. Early in the 1970s, it established, for this reason, the Hitachi Research Institute for strategic planning, enveloping the entire Hitachi group, the Systems Engineering Division to organize and coordinate interrelated activities of various business divisions, and the Systems Development Laboratory to explore new systems and solve complex problems which are peculiar to systems development. These are shown in Figure 3.

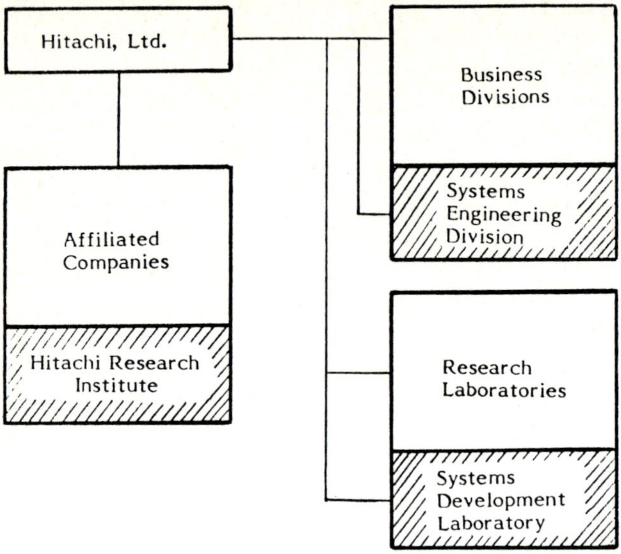

Figure 3
Organizational Structure for Systems Integration in Hitachi, Ltd.

Seen from the standpoint of policy guidance as illustrated in Figure 2, it is clear that the role of the Hitachi Research Institute corresponds to "policy formation (the third initiative)", the Systems Engineering Division facilitates "strategic coordination (the fourth initiative)", and the Systems Development Laboratory (SDL) participates in the systems development in the area of "operational experiments (the fifth initiative)".

In establishing their new roles and their relationships with other corporate activities, all three organizations face the problem of whether they will attempt to deal with the full complexity of the issue or will seek only modest incremental and low-profiled initiatives. While they appear equally to be taking a low-profiled strategy, the uniqueness of SDL is its emphasis on the development of technologies and methodologies which are accumulatable and transferrable. In order to solve systems-oriented problems for the operation level managers and professionals, the researchers at SDL have had to integrate a wide variety of technologies as well as to develop new systems technologies. The accumulated experiences of SDL at these operation level activities are becoming a source of valuable information which can now be fed back to strategy level planners as guidance.

6.2 SDL's Roles in Problem Definition (Figures 4 and 5)

The first problem in new systems exploration lies in the complexity and ambiguity of the problems themselves. From the standpoint of the manufacturer who supplies the hardware systems, the systems planners are often embarrassed by the lack of clear definition of requirements. PPDS (Planning Procedure to Develop Systems) is a representative methodology developed by SDL and furnishes the fundamental procedures and computer-aided tools to support system requirements analysis by using a structural modeling method. [9,10] Dr. K. Haruna et al. at SDL have found that the requirements analysis process consists of three unique but interdependent phases: needs analysis, functions analysis and operations analysis. They also found that planning procedures and tools could be useful only when these were compatible with actual planning processes.

The outcome of the first phase is an objectives tree. Before the construction of the objectives tree, the systems planner should gather and analyze the needs and problems of the customer. This task is, however, not easy because needs are diversified and problems are complicated. To simplify this task, PPDS includes an interest group-functions matrix method. An objectives tree, thus established, is a systematic expression of customer norms in a specific situation which is the starting point of systems development and used as the crieteria during the development process.

Secondly, the purpose of the functional requirements analysis is to find a matching between the objectives of the organization and the performance of a new system. To expedite this process, PPDS furnishes an effective methodology using functions tree and an objectives-functions interactive matrix. The functions tree is an expression of strategies to satisfy the objectives of the customer and, at the same time, an expression of goals to be pursued by the

Steps / Phases	Problem Definition and Value System Design	Systems Synthesis, Analysis and Design	Evaluation and Planning for Action
Systems Exploration	SDL assists the customer	SDL assists Hitachi sales engineering group.	SDL assists the customer
Program and Project Planning	SDL assists Hitachi engineering group.	Mainly by Hitachi engineering group.	Mainly by the customer
Implementation	Minor problems only	SDL assists Hitachi engineering group.	Minor problems only
Operation			SDL assists operation and analyzes results.

Unique contribution by SDL

Auxiliary contribution by SDL

Figure 4
The Roles of SDL in Systems Development

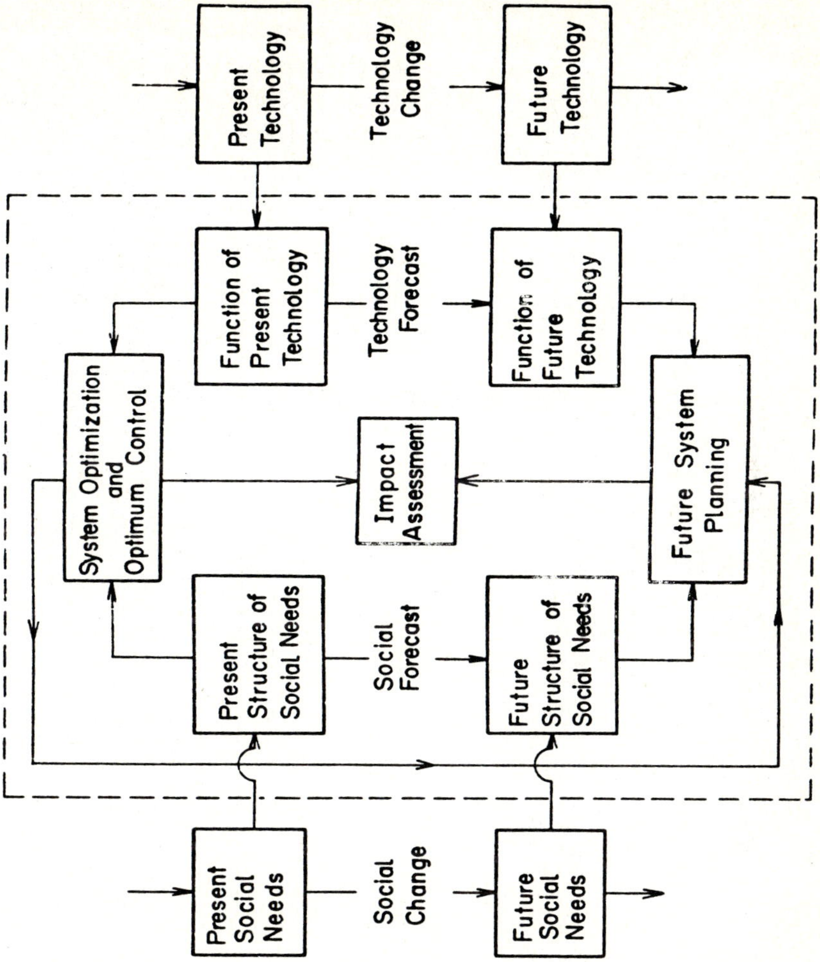

Figure 5
SDL: New Tools for Technical Direction

actual operation of a new system.

Finally, an operational activity flow is developed. It represents activities of the new system in the actual organizational situation. In determining detailed specifications and allocating limited resources, the systems planners can effectively utilize the functions and objectives trees as reliable guidlines.

SDL, in cooperation with business divisions and customers, has used PPDS to define various kinds of real but ambiguous problems. For example:

Structuring of planning models for new transit systems

System requirements analysis of computer-based informaion systems

Strategic product planning

The fundamental framework of PPDS consisting of three phases — needs, functions and operations —— is not peculiar to PPDS but universally perceived and expressed differently as follows:

"A"	"B"	"C"	"D"
needs	policy	why	objective
functions	strategy	what	goal
operations	tactics	how	action

Similar expressions and frameworks are noticed in SDL almost every day. As a consequence, many systems technologies developed by SDL for planning and evaluation are fundamentally based on the same conceptual framework.

For example, TRANSPLAN (Transportation Planning System) has been developed by SDL as a flexible hierarchical computer-aided planning system, which consists of HOPPS (Hierarchy Organization System for Urban Transportation Planning and Predict Strategies), STEPS

(Synthetic Transit Systems Evaluation and Planning Systems), and JUMPS (Justified Models for Practical Specification). First, HOPPS evaluates plans based on the social cost/benefit analysis. Next, STEPS determines the basic system specifications to minimize the construction and operation costs of the operator. Finally, JUMPS proposes detailed specifications based on a quick and accurate simulation of system performance.

In defining the requirements of users and exploring new systems, the systems planners should also consider future transitions of user needs and technologies, at least to some extent. Thus, the next role of SDL is to help planners forecast social and technological change. BEACON (Business Effective Administration and Control Support System) is a typical computer-aided tool for this purpose. BEACON consists of several functions, such as demand forecasting, market index monitoring, master production planning, inventory control, and data management. In order to remove the uncertainties and grasp the changes in demand trends, a new type of pattern recognition method has been developed and applied in BEACON.

6.3 SDL's Roles in Communication and Coordination (Figure 6)

Systems development requires a high degree of cooperation between various organizations. The next role of SDL is to facilitate this cooperation by furnishing new catalytic mechanisms for communication and coordination. If asked "what prompted the creation of a new research organization like SDL", the answer would be that the distance between potential end-user needs and potential technical system suppliers had become so large that SDL was established to bridge the gap. I would now like to further expand upon this.

To begin with, I would like to comment on the relation between SDL and the potential end-user in the systems exploration phase. The mission analysis in this phase has until recently been performed by the end-user. This leading role by systems users will not change in the foreseeable future. In the case of small and simple systems, the user can analyze and determine the mission of the system with some assistance from the sales group of the supplier. As the system becomes more complex and innovative, systematic and continued

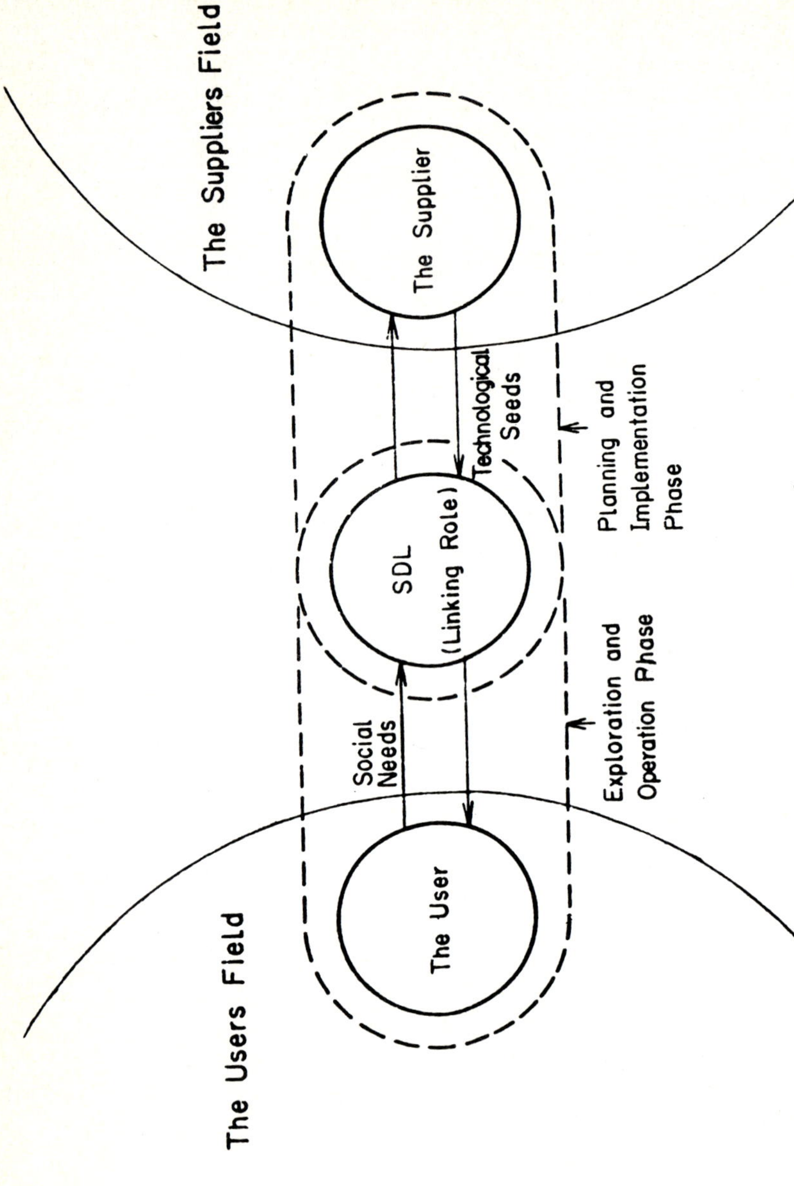

Figure 6
The Roles of SDL for Communication and Coordination

cooperation by the engineers of the supplier becomes necessary. However, because the mission analysis in the exploration phase is user-need-oriented and very dynamic, contemporary engineering groups of the suppliers, who are discipline- and product-oriented, cannot adequately respond to the requirements of the potential end-user. SDL is a new organization which is expected to lubricate this mechanism.

The second issue arising from the creation of SDL is its relation to "think tanks" or contracted research organizations, which specialize in solving the above difficulties. Because the processes of differentiation, specialization and the practice of cooperation between systems engineering and product engineering are not yet advanced in Japan, contracted research organizations independent from hardware manufacturers often experience difficulty in achieving an actually realizabale plan. In this unique Japanese situation, SDL as a research arm of a manufacturing company can achieve far more than independently contracted research organizations.

The third point is SDL's relation with development and design engineers in business divisions within the company. These engineers have been and will continue to be directly responsible for product and technology development. However, they are experiencing difficulties nowadays in developing truly market-oriented products because the market requirements often exceed their discipline-oriented approach. SDL can assist factory engineers in making up for this shortage.

In addition to the tools for problem definition which were described in the preceding section, SDL has developed other tools for communication.

The most interesting tool for communication is a know-how transfer method which has been applied to a risk management information system for large construction projects. The emphasis of this system is on how knowledge and experience obtained from one project can be applied to other projects, and how this knowledge and external factors are connected for systematic extraction.

MPSS (Management Planning Support System) is an example of tools for managing coordination developed by SDL. MPSS supports planners by expediting the evaluation of alternative strategies and the search for a better solution.

6.4 SDL's Roles in Decision-Making

One of the serious concerns of SDL is its specialty or lack of a specialty in solving systems-oriented problems, which are invariably complex, involve conflicts, contain much uncertainty, and require expertise from a number of disparate disciplines for resolution. Thus, it is not surprising that the intuitive judgements of professionals and the bargaining between managers form the usual basis for decision-making in complex organizations.

However, managers and professionals are not always satisfied with the intuitive method and seek the assistance of SDL. They expect a highly analytical and quantitative scientific approach which appears to yield much more rational and objective resolutions. The primary danger with this kind of approach is, however, that it sometimes yields sophisticated solutions to nonexistent problems.

One of the key research approaches at SDL is the integration of intuitive judgements and the analytical method. An ambitious activity along this line is the improvement of decision support systems based on artificial intelligence and natural language processing, which are expected to become key technologies in systems development.

EXCEED (Executive Management Decision Support System) was developed by the Hitachi Software Works in collaboration with SDL. The basic design emphasis of EXCEED is twofold: to implement an end-user-oriented system and an effective tool to support the developments of application programs. In order to realize this objective, it is essential that the system reflect the manner in which its users normally carry out their work. Therefore, the basic data processing cycle of EXCEED is different from the traditional cycle which receives data from transaction file, updates the master file, and finally generates reports. On the other hand, the data processing cycle of EXCEED is a natural extension of office work; that is, users get documents from their cabinet files and manually update them wherever they need to record anything.[11]

In addition to upgrading decision support systems, the continuing efforts by SDL to manage and integrate its own various functions of problem definition, communication, coordination and decision support are enhancing the effectiveness of SDL and establishing the unique domain of SDL. Thus, SDL is becoming an exploratory center for systems development and serves as a neutral meeting ground for managers and professionals from various parts of the business divisions as well as a meeting ground for suppliers and users. In other words, SDL is a pilot model of PGS for complex innovations.

7 CONCLUDING REMARKS

It is still too early to evaluate or predict the effectiveness of PGS based on pioneering activities in Japan such as SDL of Hitachi, Ltd. However, these activities are already providing useful contributions to the ongoing company processes, and it is expected that the overall impact of PGS over time will be enormous. The short-term goal of PGS is to compensate for the limitations of existing management systems. Its long-term contribution will be as a catalyst for organizational innovation in other parts of the company grappling with the enhancement of "corporate capability for managing change". [12]

As it develops in this direction, PGS will be put to many tests, be improved because of these experiences and propagate the lessons learned to the company and society at large. "Gaming Simulation" as a future language will play an important role in the process. [13]

ACKNOWLEDGEMENTS

The author wishes to express his gratitude to Dr. T. Miura of

Systems Development Laboratory for his guidance and encouragement and to Drs. K. Haruna, K. Sato and K. Nakao for their valuable suggestions.

REFERENCES:

[1] Puri, T. and Bhide, A., The Crucial Weaknesses of Japan Inc., the Wall Street Journal, June 8, 1981

[2] Drucker, P. F., Behind Japan's Success, Harvard Business Review 59(1) 83-90 (Jan.-Feb., 1981)

[3] Yergin, D. and Hillenbrand, M., eds., Global Insecurity: A Strategy for Energy and Economic Renewal, Houghton Mifflin, Boston, 1982

[4] Williams, C. et al., Planning and Evaluation Assistance to the NSF Experimental R&D Incentives Program, NSF Contract URU-72-66, SRI Project 2291, Stanford Research Institute, Menlo Park, June 1973

[5] Watson, R. H., Interpretive Structural Modeling ——A Useful Tool for Technology Assessment?, Technology Frecasting and Social Change, 11(2) 165-185 (1978)

[6] Bennis, W.G. et al. (eds.), The Planning of Change, Holt, Rinehart and Winston, New York, 1969

[7] Wheelwright, S.C., Japan Where Operations Really Are Strategic, Harvard Business Review 59(4) 67-74 (July-Aug., 1981)

[8] Shibata, Y., Remarks on the Role of Technology Assessment, in: Boroush, M.A. et al. (eds.), Technology Assessment: Creative Futures, North Holland, New York, 1980

[9] Nakao, K. et al., A Structural Approach to System Requirements

Analysis of Information Systems, Proc. of COMPSAC, 207-213 (1980)

[10] Komoda, N. et al., An Innovative Approach to System Requirements Analysis by Using Structural Modeling Method, Proc. of Conference on Software Engineering, 305-313 (1981)

[11] Isobe, H. et al., EXCEED —— Excutive Decision Support System, Proc. of 1982 APL Users Meeting, vol.1, 296-324 (1982)

[12] Ansoff, H. Igor et al. (eds.), From Strategic Planning to Strategic Management, John Wiley & Sons, New York, 1976

[13] Duke, R. D., Gaming: The Future's Language, Sage Publications, New York, 1974

R & D Managment Systems in Japanese Industry
H. Eto and K. Matsui (editors)
© Elsevier Science Publishers B.V. (North-Holland), 1984

Chapter III PROBLEM OF PRESENT EVALUATION OF PUBLIC INVESTMENT AND PROPOSAL OF NEW EVALUATION OF R & D INVESTMENT AS REFLATIONARY POLICY

SHOGO SAKAKURA

Director, Office for Moonlight Project
Agency of Industrial Science & Technology
Ministry of International Trade & Industry
Tokyo, Japan

Nowadays, public investment is used to stimulate the economic market and employment. But this public investment faces problems. A large portion of public investment is spent for the acquisition of land which has little stimulation on the national economy. Public investment of construction tends to have long delays due to resident opposition and problems of land acquisition. Because of these problems, there are many arguments that effectiveness of public investment as a stimulator for the national economy is becoming less and less.

Here the author would like to point out the importance of R&D investment by government. Considering the short range behavior of expenditure, since R&D investment is not different from public investment in use, it has enough economic market and employment stimulation effects. Moreover, it does not have expense for land acquision and there is no delay in spending. Therefore, effects of R&D investment as a stimulator for the national economy are larger than public investment.

1. INTRODUCTION

After World War II, Japan adopted, digested, and further developed necessary technologies originally developed in Euro-American industrialized countries, and achieved a high rate of growth with those technologies as stepping stones.

However, this pattern came to an end by 1970. The level of technology of Japanese industries in many fields matched and surpassed those of advanced industrialized countries and it became very difficult to buy and introduce new foreign technologies in Japan. Besides, technology has become an important weapon in international politics and trade negotiations, as represented by the recent US-Japan high technology disputes. Forein pioneering technologies are unable to be imported without cross-licensing or unless there is joint development.

As Japan is very poor in natural resources such as food, energy and others, she must import these resources while exporting her excellent industrial products to accumulate foreign currency. Efforts have started in Japan, to develop original technologies under the slogan of "a nation based on technology."

Individual companies are making every effort with the same view point, however Japanese companies have little experience with large R&D investment and have a poor funding capability. M.I.T.I. assists private industry's R&D using consigned research or direct aid in the case of big projects. Examples are the "sun-shine project" (new energy technology development projects such as geothermal power generation, solar cells), "moon-light project" (energy saving technology development projects such as fuel cells, high efficiency gas turbines), "basic technology development of next generation industries", various development aid to important R&D (super LSI, aircrafts, jet-engines, computers, etc.). Although it is generally said in the United States that only one or two R&D projects out of ten succeed, in Japan, since M.I.T.I. often selects mainly those projects which seem to be able to succeed for development aid, several projects out of ten show some successful results. Examples of large success are the development of super LSI and computers. (It is the author's opinion that M.I.T.I. should accept R&D projects with greater development risk for aid, however budget limitations force M.I.T.I. to minimize the risks and adopt a conservative funding policy.)

The other day, the author had the opportunity to visit Toshiba Total Technology Exibition (Toshiba Sogo Gijitsu Ten) which was held in Hotel New Ootani. Among exhibited there were state of the art technologies under development in cooperation with M.I.T.I. such as gas turbines, fuel cells, DC transmission, inspection robots, automatic character readers, vocal wordprocessors, optical semiconductor lasers, vertically magnitized VTR tape, VLSI manufacturing equipments, micro-processors, GaAs super high speed logical units, etc.. It was encouraging for the author to see that M.I.T.I. is fostering the seeds of technologies which support Japan's future.

Japan's R&D policy however, is now facing a great turning point. For, in the current storm of financial rebuilding and administrative reform, M.I.T.I. is experiencing severe shortages of R&D budget funds. Furthermore they have been asked to reduce the amount of aid to large companies. It is a lamentable situation that M.I.T.I. had to accept a 1983 budget allotment almost equal to that of the 1982 budget. If this trend continues, there is almost no chance of recovery after 1984, and M.I.T.I. will surely be driven out of existance.

If the author is asked the necessity of M.I.T.I.-led-R&D in the coming years, the answer is certainly affirmative. In the present long lasting "depression", private companies are slimming their budget by reducing first R&D fund. This fund in fact is based on their long-range view but will not yield results in the short run. Besides, technologies which must be developed in the coming years are becoming more and more sophisticated, and require large-scale R&D. On the other hand, the so-called developmetn risks are increasing rapidly. Considering the view of trade conflict avoidance and promotion of industrial cooperation, Japan should make a continuous effort to produce products using new technologies, export them, and at some phase develop joint production ventures while continuing to develop new technologies in Japan. Therefore, M.I.T.I's role in the field of R&D will grow larger and larger.

2. BUDGET FOR R&D

The Ministry of Finance takes the position that there is no countermeasure due to lack of financial resources. Is it true that there are no financial resources? The author's view is there are sufficient resources. The budget bill relating to promotion of science and technology is about 400 billion yen and the public investment budget in general account amounts to 6 trillion 700 billion yen or about 17 times that related to science and technology promotion. Therefore, it is possible to secure enough financial resources for R&D by utilizing public investment monies, and to drastically increase the R&D investment budget.

There may be some objection that public investment and R&D investment cannot be compared in the same dimension because objectives of the two are different. One of the largest objectives of public investment is in the stimulation of economic activity (reflationary policy). From this standpoint, R&D investment is not different in its way of the usage and, looked at from short range expenditure, it has enough market and employment stimulation effect, although R&D will not necessarily succeed. Furthermore, in the long run, both can be considered the same that they both provide a basis for the nation's standard of living.

There is also the view regarding public aid of private R&D investment. Private companies which receive aid get large profits in the case of successful technological development, and therefore it is not appropriate to spend taxes from the national budget for a few private companies.

When they succeed, however, it is possible to recover enough investment return from tax revenue. Tax revenues several times that of the national aid are returned due to multiplier effects which is discussed later, in addition to corporate taxes and fixed property taxes, and income taxes from the employees of these companies.

3. NEW EVALUATION OF R&D INVESTMENT EFFECTS

Because there are many studies being done on the effects of R&D investment on economic growth, two calculations can be derived supporting the above view point, as shown below.

3.1 Market Stimulation Effect (Reflationary Effect)

The first is the market stimulation effect during depression. Public investment brings an increase in the income of related companies and their employees in the form of profits and wages through the project by an equal amount to the investment. This income increase yields a comsupmption increase by the amount times marginal propensity to consume. This in turn becomes the next income increase and which again yields consumption. In the same manner, the income increase due to the original investment brings incomes consecutively through the increase of consumption as the medium. This is the so-called Keynesian investment multiplier effect, and the total of these derived incomes divided by the amount of the original investment is the investment multiplier. Setting aside the view that the Keynesian theory is no longer effective, it is for this effect that public investment is used for stimulation of economic activity.

In this sense, R&D investment can play the same role as the public investment. R&D investment, (the weight of expenditure items differs slightly from public investment) consists of wages, equipment costs, costs of materials and parts, utilities, etc.. These expenditures bring an income increase in R&D related sectors, which are spent for consumption, and that in turn yields the next income, which is exactly the same as the multiplier effect of public investment.

Further, the funds for public investment are raised mainly by issuing public bonds which equal the total of savings accumulated in the multiplier effect process, and it appears as if these savings are absorbed by the issuance of public bonds. R&D investment can be considered in the same way if the funds raised by issuance of public bonds are invested, because it would exactly balance with the saving accumulated from the multiplier process.

What the author wants to emphasize here in the discussion of multiplier effects is R&D investment has larger effects than public investment. In the case of roads, railroads, airports, and city developments, etc, which consist the major part of public investment, the proportion spent for land acquisition is very large and becomes income of the land owners. However, the propensity of the land owners to save is usually very large compared with that of the general public in the multiplier effect process.

Let us consider the above using figures. Let's set the amount of public investment as P, the proportion of land acquisition as l, the propensity of land owners to save as and the propensity of general public to save as i_g.

The income increase yielded by the original public investment is P. The consumption (or next income) Q brought by this P becomes"

$$Q = Pl(1-i_l) + P(1-l)(1-i_g)$$

The next consumption (or next income) yielded by this Q is $Q(1-i_g)$ which yields the incomes $Q(1-i_g)^2$, $Q(1-i_g)^3$ ...consecutively.

The sum of the incomes becomes:

$$P + Q\{1+(1-i_g) + (1-i_g)^2 + (1-i_g)^3 + \ldots\}$$

$$= P + Q\left\{\frac{1}{1-(1-i_g)}\right\} = P + Q\cdot\frac{1}{i_g}$$

$$= P + \{ Pl(1-i_l) + P(1-l)(1-i_g)\}\cdot\frac{1}{i_g}$$

$$P\left\{\frac{l- (i_l-i_g)}{i_g}\right\}$$

The investment multiplier of the public investment K_P is derived dividing this by P:

$$K_P = \frac{1-l(i_l-i_g)}{i_g}$$

Modifying this by setting the proportion of public investment supplied by import as r_p and the proportion of consumption derived by multiplier effects supplied by import as r_g, we get:

$$K_p = (1-r_p)\frac{\{1-l(1-r_g)(i_l-i_g)\}}{i_g+r_g-i_g\cdot r_g}$$

On the other hand, the investment multiplier of R&D investment K_R is,

$$K_R = \frac{1}{i_g}$$

because l equals zero. Modifying it, putting the proportion of R&D investment supplied by import as r_R, we get:

$$K_R = (1-r_R)\frac{1}{i_g+r_g-i_g\cdot r_g}$$

Such modification on import is not necessary in a simple comparison of public investment and R&D investment, but in the calculation of tax revenue as will be stated later, we over-estimate without this modification and is shown here.

Putting aside this import modification, we see clearly from K_P and K_R what has been stated before. That is, the investment multiplier of public investment is smaller than that of R&D investment by the amount shown in the next formula.

$$\left(\frac{\text{the Proportion of land acquisition}}{\text{the propensity to save by the general public}}\right) \times \text{(the propensity to}$$
save by land owners- propensity to save by the general public)

We can see from this formula that the higher the proportion of land acquisition in the public investment and the broader the difference of the propensity to save between the land owners and general public, the smaller the investment multiplier of the public investment.

The proportion of land acquisition of current public investment is about 25%. Regarding the propensity to save by land owners as about 70% and the general public as about 20%, we get $K_P = 4.38$, $K_R = 5$, and K_P is smaller than K_R by about 12%.

Another problem with public investment is the delay in construction works. The multiplier discussed above assumes that construction works are done as scheduled. However, in recent public investment, there are frequent delays due to resident opposition and problems of acquisition of construction site. Construction is often shelved as long as two or three years after the allotment of budget. In these cases, far from the multiplier effects, even the first expenditure is not spent. On the other hand, in the case of R&D investment, there are not such delays. As such, although nobody has indicated so far, R&D investment has far larger reflationary effects and accordingly considerably larger market and employment stimulation effect than public investment.

The above discussion has been based on the Keynesian multiplier. The "multiplier effect" is also used to mean the production inducement coefficient (out-put coefficient) in the input-output analysis which is obtained by putting the public investment in the final demand, computing the intermediate production induced, and dividing the sum of the final demand and the intermediate production by the original public investment. The expression of the multiplier effect as 1.6 or 1.8 usually means the coefficient is obtained as such. Looking from the production inducement effect, public investment is apparantly smaller than R&D investment because the proportion of land acquisition is larger.

This calculation does not contain, different from caliculation of the Keynesian multiplier, the process in which the original investment yields income, in turn increases consumption and yields income again, and under-estimates the effects of public investment (which is the same in the case of R&D investment). On the contrary, calculation of the Keynesian multiplier only considers final demands and does not contain production inducement effects as in the input-output analysis. Therefore, properly speaking, a combined analysis is needed. This is not discussed in this paper due to page constraint.

3.2 Tax Revenues Gained through R&D

The second is the comparison between government aid of R&D and the tax revenues obtained from successful R&D.

R&D does not necessarily succeed, and rather in fact, there are more unsuccessful cases. The companies which receive R&D consignments and government aid do not have any special profit in the case of failure. While, in successful cases the result of R&D results in a companies' new product and increases sales. Depending on the amount of sales increase, this often brings continuous long lasting profit to the company.

Looking at this point only, the author agrees with the view stated before that it is not appropriate to spend national budge moneies collected through taxes for one company. This point differs largely from public investment.

However, these companies pay corporate taxes from profits gained through sales increases of new products developed by the success of R&D. They also pay fixed property taxes because they invest in production facilities necessary for production. The company pays dividends to its stockholders and bonuses to its exective directors from the profits. Those stockholders and directors in turn pay income taxes and resident taxes. The number of employees naturally increases and these employees pay income taxes and resident taxes from salaries received in their jobs which are related to the new products.

Further, this company will purchase, because of the new products, energy, materials, parts, etc., and additionaly spend money on equipment, and pay interest for borrowing necessary funds from banks. Accordingly, sales of the suppliers of energy, materials, parts, equipment etc. will increase, while income of the banks increases as well. Sales of various service companies related to the new products, such as service, transportation, etc, will also increase.

The sales increase of these related companies brings the national government and local government corporate taxes and fixed property taxes, as well as income taxes and resident taxes from stockholders, directors, and employees. Further, if these companies purchase goods and services relating to sales increases, there will be various tax increases at every step until it reaches the stage of importing from abroad. Since this continues until the new product ends its life-cycle, the sum of the tax revenues reaches a considerable amount.

The above discussion focused only on tax revenues related to the new product, and in addition to this, income increases related to sales increases of the new product at every step stated above increase consumption and yield additional income from which there will be a tax increase. Further this income in turn generates greater consumption that will again yield income and income taxes. This is, so to speak, a multiplier effect of tax revenues. Since there is no such tax revenue increase in absence of the original government R&D aid, it is proper to include such tax revenues as a return on the aid.

Discussed below is the calculation method of the tax revenue increase related to the new product and those from its multiplier effects under certain assumptions.

Tax Revenues Related to the New Product

First is tax revenue related to the new product. From the sales of the new product, profit is gained, which is divided into those moneies paid to the national and local governments in the form of corporate taxes, those paid to the stockholders as dividends, those paid to exective directors as bonuses, and those saved within the corporation as reserved profit. Dividends and bonuses become the personal income of the stockholders and directors, from which income taxes and resident taxes are paid. As to the dividends of corporate stockholders, it becomes profit of these companies and is taxed any way, and it need not be given any special treatment of calculation here. Reserved profit is also disregarded, since it will be finally treated as dividends and bonuses.

Discused next are items of expenditure such as costs related to sales. The simplest is the fixed property taxes. It becomes tax revenue as local taxes. Labor cost of the employees becomes personal income from which income taxes and resident taxes are paid. Next are goods purchased from other companies, which consist of "wear and tear" goods, parts, energy like electricity and gas, etc. All these become sales of various companies in each industry. Depreciation as to machines-and-equipments and buildings are summed up as sales of other industries because it is possible to suppose that those machines-and-equipments and buildings are purchased in the end on installment equal to durable years from these industries. There are purchases of other services such as transportation, etc., but these become sales of service companies. As to interest, it becomes income (sales) of banks and insurance companies. Interest of corporate bonds held by individuals becomes personal income, but it is not necessary to itemize it here.

Discussed above were purchases of goods and services of the company which is selling the new product, and all become sales of companies in other industries. From these sales are paid the various tazes as discussed above. They will yield second and third sales of goods and services of compnaies in other industries. This process continues until it reaches goods and services which are imported from abroad and stops.

Let us consider sales of the new product and related purchases of goods and services as a whole. Let's assume the sales of the new product as A hundred million yen a year.

We denote by α_0 the proportion in A of the total of the corporate taxes and the fixed property taxes of the company which sells the new product, the companies from which the goods and services are purchased, next companies from which next goods and services are purchased, and so forth.

Next we put the ration of the income taxes and resident taxes in A as β_0. If we further denote the proportion in A of materials such as final imported energy and raw materials as r_0, and the combined marginal rate of income taxes and resident taxes in personal income as ρ, the sum of personal incomes such as stockholders' dividends, directors' bonuses, employees' salaries of all companies which started selling the new product, companies supplying goods and services to it, and the next group of companies supplying goods and services to them, and so forth is obtained by subtracting corporate taxes and fixed property taxes, and costs of imported energy and raw materials from all sales, and becomes $A\cdot(1-\alpha_0-r_0)$. Therefore, the total of income taxes and resident taxes obtained from it is $A\cdot\rho\cdot(1-\alpha_0-r_0)$. From this β_0 is obtained, $\beta_0=\rho\cdot(1-\alpha_0-r_0)$. The sum of the tax revenues collected from A becomes $A\cdot(\alpha_0+\beta_0)$. In reality, sales of the new product will follow so to speak, a lifecycle starting with a very small amount, increasing gradually, reaching a peak at some stage, and decreasing gradually. But, here, we consider it on average for simplicity, assuming that sales of A hundred million yen will continue N years. The tax revenues during this life cycle period becomes $N\cdot A\cdot(\alpha_0+\beta_0)$.

Tax Revenues due to the Multiplier Effect of New Product Sales

Discussed next is the calculation of tax revenues generated by the multiplier effect of the new product sales.

We denote the proportion of the sum of the corporate taxes and the fixed property taxes in sales (=consumptions) due to the multiplier effect as α_g, the proportion of the sum of the income taxes and resident taxes of individuals in the same sales as β_g, and the proportion of the average cost of materials such as imported energy and raw materials as r_g which is stated in 3 (a). We adopt as the marginal rate of income taxes and inhabitant taxes the same ρ as before.

Then we get:

$$\beta_g = \rho\cdot(1-\alpha_g-r_g)$$

If we put the marginal propensity to save as i_g as discussed before, the consumption (= the sales of the next round) increased by the income yielded by the A hundred million yen of new product sales is obtained subtracting outflow abroad and savings from sales, and we get:

$$A\cdot(1-r_0)(1-i_g) = A\cdot(1-r_0-i_g+r_0\cdot i_g)$$

The sales yielded by consumption at the next stage is what is obtained subtracting outflow abroad and savings from $A\cdot(1-r_0-i_g+r_0\cdot i_g)$, and we get:

$$A\cdot(1-r_0-i_g+r_0\cdot i_g)(1-r_g)(1-i_g)$$
$$= A\cdot(1-r_0-i_0+r_0\cdot i_g)(1-r_g-i_g+r_g\cdot i_g)$$

The total of sales yielded consecutively by such multiplier effect is,

$$A\cdot(1-r_0-i_g+r_0\cdot i_g)/\{1-(1-r_g-i_g+r_g\cdot i_g)\}$$
$$= A\cdot(1-r_0-i_g+r_0\cdot i_g)/(r_g+i_g-r_g\cdot i_g)$$

The total tax revenues obtained from it in N years is given as follows.

$$N\cdot A\cdot(\alpha_g+\beta_g)(1-r_0-i_g+r_0\cdot i_g)/(r_g+i_g-r_g\cdot i_g)$$

If we compare the tax revenues computed as such with the sum of R&D aid and consignment from the national government, we see the tax revenue return. (In actual computation, it must be adjusted by further discounting to the present value or by taking delay of the multiplier effect into consideration.)

Now, let us calculate a case under bold assumptions on each coefficients as an example. The sum of M.I.T.I's development aid to new computer R&D over five years from 1972 to 76 was about 70 billion yen, and the sale of new computers developed by the project was, assuming the lifecycle is 5 years, about 320 billion yen (estimate) a year on average. If we put,

$\alpha_0 = 0.04$, $r_0 = 0.1$, $\rho = 0.15$, we get

$$\beta_0 = \rho\cdot(1-\alpha_0-r_0) = 0.15\ (1-0.04-0.10) = 0.129$$

$$N\cdot A\cdot(\alpha_0+\beta_0) = 5 \times 320 \times (0.04 + 0.129) = 270.4$$

The tax revenues during the five years related to the new product is 270.4 billion yen.

Further, tax revenues due to the multiplier effect of the sale of the new product is, putting,

$$\alpha_g = 0.02,\ r_g = 0.05,\ i_g = 0.2,$$

$$\beta_g = \rho(1-\alpha_g-r_g) = 0.15(1-0.02-0.05) = 0.14$$

$$N\cdot A\cdot(\alpha_g + \beta_g)(1-r_0-i_g+r_0\cdot i_g)(r_g+i_g-r_g\cdot i_g)$$

$$= 5 \times 320\ (0.02+0.14)(1-0.1-0.2+0.02)/(0.05+0.2-0.01)$$

$$= 5 \times 320 \times 0.16 \times 0.72 \times 0.24 = 768$$

768 billion yen is obtained over 5 years. The total of tax revenue is 1 trillion 38.4 billion yen, and so with government R&D aid of about 70 billion, about 15 times more tax revenue is obtained.

In this case the portion of government R&D aid of total R&D cost was 50%, and therefore, the total R&D cost was 140 billion yen. The average sale per year increased by 320 billion yen, and the figure indicates sales gains by research costs were rather small. Usually, the amount of new product sales measured against research cost is larger than this, and a number (15 times) will be far greater.

4. CONSLUSION

Some people believe it is not appropriate to spend taxes for R&D investment in private companies, as in cases where there is successful technological development, these companies will receive large profits. However, this belief is not correct, because when these companies succeed, tax revenue as a whole increases several times that of the government aid, when one takes into account total tax revenue such as corporate taxes, fixed property taxes and income taxes of employees of these companies. And also in the broader sense, there are multiple effects in tax revenue as a whole.

As was stated here, if a company receives from the government R&D aid and succeeds in developing a new product, it generates tax revenues of at least 10 times that of the aid received to the national treasury. The above discussion was limited to the direct return in the form of tax revenues from the R&D investment. It can be easily understood, as one can see in the role played by R&D on the changes of industrial structure, that there is great effect of various R&D, although we cannot compute in regard to the productivity increase of other industries and the improvement of the nation's living standards.

When the rate of success of government aided R&D is low, expenses and income cannot balance. But in Japan's case, since more than 50% of the projects will succeed, it can be said that, though there is a question of timing, tax revenues which are several times more than the government's R&D expenditure are gained in the end. If one correctly recognizes this meaning of R&D, it can be said that what is really necessary for Japan now is a drastic increase of R&D investment for the stimulation of economic activities and we should evaluate R&D investment as a reflationary policy.

R & D Managment Systems in Japanese Industry
H. Eto and K. Matsui (editors)
© Elsevier Science Publishers B.V. (North-Holland), 1984

Chapter IV ON JAPANESE CREATIVITY

ANDERS DIEHL

Counsellor Science and Technology
Embassy of Finland
Tokyo, Japan

In Japan both the foreigners and the Japanese are actively debating over, whether Japan, the Japanese people or the Japanese system is creative. Until now the Japanese have undoubtedly contributed to the development of a new basic innovation on a little scale. The author believes this does not result from a limited creativity in the Japanese as compared to the others, but is due to the fact that there has been no great need or motivation for this kind of creativity. Now the situtation is changing and the focus is directed towards creativity. Both on the government and industry level creativity is actively encouraged and developed. The author sees a possible parallel to the development of Quality Control. In 30 years this completely changed the quality of Japanese products and companies. With a similar effort the creativity will possibly be developed to the extent that we in 20 - 30 years will wake up to find that the Japanese have surpassed most western nations in creativity: the same way we woke up to find that in quality the Japanese had surpassed the West in 30 years.

1 JAPANESE CREATIVITY?

"Creativity", "innovation" and other similar words are very popular topics in Japan or in discussions about Japan. Both the Japanese and foreigners are discussing about Japan from a viewpoint that Japan and the Japanese people have not shown particular "creativity" in the past and should now develop this property. Implied in these discussions is that other nations or nationalities have shown this "creativity", but the Japanese have until now shown only a very good capability of improving what the others have created and of massproducing it.

Why have the Japanese earned this reputation of not being creative?

It is evident that most of the products with which the Japanese industry today is conquering the world are improvements of products that are invented elsewhere. But is this a proof of the Japanese not being creative? Is it not just a proof of a sound industrial policy? The Japanese had to build up a solid industrial base from the destruction of the second world war. The nation rallied around a common objective to reach the standard of living, whatever that is, of the western industrialized countries. The best policy to reach this objective was to learn from these countries and do, what they had done, improving what could be improved. There was no need to or reason to re-invent gunpowder or any other product.

There were no objections in Japan either against learning, even from previous enemies. The country has a long history of learning from other countries and adapting foreign thoughts or methods to its own needs. Japan has learnt in turns from India, China, Great Britain, France, Germany, U.S.A. and others.

The Japanese also have a great enthusiasm and even pride of learning from others. It is surprising how often they themselves point out that something is learnt from abroad. In many countries there is a tendency to exaggerate, what is invented in the own country, but sometimes you can get the feeling that the Japanese modesty exaggerates the role of foreigners. For example the Japanese will often point out that the Japanese writing is imported from China and take that as a fact suggesting lack of creativity in Japan. But not American or European would suggest that importing the alphabet has been a proof of lack of creativity in those countries. Still the Japanese developed their own phonetic text based on the Chinese characters.

On another level the Japanese companies do not have the mentality that all basic technology should be developed by themselves: only the technology that cannot be bought must be developed. The companies even proudly announced in their annual reports that they bought a license from some company instead of silently being ashamed of the fact that their own engineers did not develop it themselves.

In addition to having a mentality that supports learning from other nations the Japanese have been formidably efficient students. In many cases they have in a surprisingly short time been able to improve the methods they have learnt from abroad and to manufacture better products than the company they learnt from.

Perhaps one of the reasons for being accused of copying is that the Japanese have been better than the others in implementing and improving what they have learned. Everybody tries the same methods, but only a few succeed.

Another reason exists for the Japanese to gain the reputation of only improving what others invent and not developing own technology. As a company policy Japanese companies very seldom sell any technology they have developed, they sell products. This easily gives the impression that they do not have technology to sell.

Official statistics try today to prove that Japan also sells technology. Overall statistics made by The Science and Technology Agency 1982: "Indicators of Science and Technology" show that the ratio between sales and purchase of technology 1980 was 0,67, but checking the details show that the ratio for developed countries like USA was 0,14 and eg. Federal Republic of Germany 0,17. These statistics further contain the "know-how exports" of Japanese companies to their subsidiaries in respective countries.

Thus the Japanese, instead of making propaganda for their own technology developments for example by trying to sell them, sell products. Companies in many other countries try actively to sell their technology, thus giving the impression they have developed much more than the others.

People both inside and outside Japan argue that the small number of Nobel laureates given to Japanese researchers present another proof of lack of creativity in Japan. But it should be remembered that these laureates are divided by European judges according to European standards.

Another striking fact is that during the last 10 years Japanese researchers have received 2 laureates, French 1 and German 3. This certainly does not prove that Japan is less creative than these advanced nations. The overwhelming majority of the Nobel laureates have been given to the researchers of the US.

When I as a young schoolboy some 30 years ago read my first book on the development of technology and industry, I remember clearly that the authors of the book stated that most of all inventions were made in European countries, but massproduced in the United States, where massproduction technology was well developed. On the contrary creativity and innovations were mostly found in Europe, but not in the US.

It seems that the situation has changed radically in 30 years: Japan has taken the role of the country for massproduction and the US the role of an innovator.

The role and image of the US has changed in 30 years. It seems to me very likely that Japan will go through the same change in some time.

2 OBSTACLES FOR CREATIVITY

In the previous section I have discussed some of the arguments used to show that Japan lacks creativity. In my opinion they just show that Japan during the period after the second world war has had other priorities than basic "creative" science. There are factors in Japanese organizations that certainly can be hampering the development of new ideas.

The strong emphasis on group work can make it difficult for individuals with original ideas to persuade the group to make consensus around his suggestions, if they are too original. Without group consensus it can be difficult to develop these ideas.

Another problem is lack of criticism of ideas established by seniors. The rigid seniority system makes this a delicate matter. In this way criticism and debate that could provoke new ideas is rather rare.

Critics agree that the Japanese way of thinking is less concrete and logical than the western way. This can also be a problem when new ideas should be developed into practical applications.

These and other examples show that the Japanese society certainly exhibits some factors which can make it difficult for creative ideas to develop. On the other hand certain restraining factors, like religion which in many cases has hampered scientific development, do not exist in Japan.

These negative factors for the development of creativity have in the same way also been obstacles for the development of the industry. The example of the "Quality Control Movement" show that this kind of obstacles can be overcome through education in a reasonably short term.

3 "QUALITY CONTROL": FOSTERING OF CREATIVITY ON A LARGE SCALE

The "Quality Control" (QC) movement started over 30 years ago in a situation where the quality of Japanese products had a solid reputation of being of inferior quality.

To improve the quality of the products it was not enough to concentrate on technical development of products and production techniques: the way of thinking of the employees and the organizations had to be changed. The same factors that perhaps are hampering the creativity today were obstacles in the factories and laboratories responsible for the quality of the products.

Three decades of massive education of all the employees in companies applying "QC" methods has changed the ways of thinking and working within the company organizations.

The QC education contains many of the elements we consider necessary for the development of creativity: isolation of problems with statistical analysis like regression or Pareto diagram, brain storming to generate ideas, cause-and-effect analysis with Ishikawa diagram etc.

This amounts to large scale creativity development program for both factory workers and other employees. Estimates vary considerably but at least several millions of employees have been educated in this program.

The results have been impressive. Through a very large number of small innovations and improvements the Japanese companies have been able to reach results that in many fields are superior to other countries. This "creativity by small improvements" should in my opinion be regarded as creative as an innovation by one large step, if the result is similar, or as often is the case, better.

It is significant that the new way of reasoning and improving development in the QC-movement is not very much used outside the sphere of influence of this movement in Japan. Outside this sphere many Japanese often lack the concreteness and logical way of thinking of westerners and this gives rise to many frustrating misunderstandings.

This interest for improvements is perhaps partly based on special Japanese features, but it has been organized and motivated through QC, which also has provided concrete methods that everybody can be used for improvements.

The Japanese mentality for improvement was pointed out by Dr. Juran, one of the founders of the QC during The International Conference on Quality Control" in Tokyo 1978.
At the conference the Japanese and foreign speakers had all presented results which factually were very similar. There was however a small but significant difference in the way of presentation: The Japanese talked about "this is the way we improve", while the others talked about "this is what we did".

The QC movement has been very successful in promoting creativity in the factories in a large scale.

There are other areas where the Japanese also have displayed creativity to a large extent. The individuals behind companies like Matsushita, Honda, Sony etc. have made a larger creative contribution to society than most other people. Their creativity has been directed to building up high-technology companies. Under other circumstances, motivating these individuals in other directions they could for example have directed their energy to promote science. There was no need or room for that in Japan, when they started their work.

4 PROMOTION OF CREATIVITY

As a result of both internal and foreign criticism of the lack of "creative research" in Japan many organizations and individuals are now actively trying to promote creativity and innovations. There is a difference in approach between Japan and other countries.

The Japanese start from the assumption that creativity is lower than elsewhere and should be improved. Most other countries are assuming their creativity is reasonably high. This assumption of being inferior motivates the Japanese to a large effort of improvement. The example of QC shows that this kind of effort can be highly successful.

The government program now contains several measures to improve creativity. Ministry of International Trade and Industry stressed in many occasions the need for creativity, when they formulated their vision for policies in 1980's. The general statements about creativity in the government policies are bound to motivate other organizations to organize promotion of creativity. On a more concrete level the Agency of Industrial Science and Technology has started specific research programs on "creative science".

Many companies have a long time ago started programs for creativity. For example Honda's "Idea contest" for promoting innovations and Mitsubishi Electrics' creativity education programs are famous.

All the medias are also often discussing creativity and how to promote it. Improvement of creativity is on the way of becoming a national consensus. This leads in a short time to a similar nationwide movement as the Quality Control.

When this has started, it is only a question of time when we see the results. Then it is not very relevant to discuss, whether the Japanese start their creativity development from a level slightly lower than most western countries or not. With continuing improvement of education and promotion they are bound to reach a very high level.

Are we going to wake up in 20 - 30 years and find that the Japanese are far ahead of other countries in creativity and innovation, in the same way as we recently woke up to find they are far ahead in quality as a result of their 30 years of Quality Control movement?

Or is it time for us to search our souls to see, if other countries really are more creative than Japan? Then we should start discussing and improving creativity in our own organizations.

R & D Managment Systems in Japanese Industry
H. Eto and K. Matsui (editors)
© Elsevier Science Publishers B.V. (North-Holland), 1984

Chapter V A FRAMEWORK FOR INVESTIGATIONS OF INFORMATION TECHNOLOGY-BASED INNOVATION

KOZO SUGIYAMA

Policy Science Unit
International Institute for Advanced Study of
Social Information Science, Fujitsu Limited
140 Miyamoto, Numazu, Shizuoka 410-03, Japan

A tentative framework for investigating "information innovation" or innovation based upon information technology is developed by searching more than two hundred literatures(mainly from Japan) from both a technological and societal point of view. In the framework the integration of control, communication and computation technologies is emphasized as a key-concept. The framework is illustrated in the form of diagrams and a thesaurus which help us present our concepts and views in a clearer and more interesting manner.

1 INTRODUCTION

The purpose of this paper is to present a framework for investigating "information innovation" or innovation based upon information technology, which is proceeding apace not only in Japan, but also in several other countries, and is becoming a major trend in the world. Before we can begin to investigate a complex problem, we need a framework for it, usually constituted of the elements of the problem and the relations among them(see B56). Of course it is dif-

---

Basic part of this research was performed at the International Institute for Advanced Study of Social Information Science, Fujitsu Ltd., and completed while the author was at the International Institute for Applied Systems Analysis, Austria. Helpful advice from Prof. Tibor Vasko, Dr. Tosio Kitagawa and Dr. Mitsuhiko Toda is gratefully achnowledged.

ficult to get a complete final picture in the early stages of research. However, even a tentative one can help us grasp an outline of the whole structure of the problem so that we can begin more detailed research with a more error-proof strategy.

In order to develop a tentative framework, we have collected more than two hundred articles (mainly from Japan) that are likely to be relevant to "information innovation", and that deal with a range of topics from hardware to social impacts. Then we have read through these articles, forming concepts and views (we were particularly influenced by articles R14, B30, B23, B56, P33, P101, P109 and P110) and illustrating them in the form of diagrams, which help us present our concepts and views in a clearer and more interesting manner and which are quite convenient for discussions.

Because this type of innovation is a social phenomenon and has great impacts upon society, we begin Section 2 by presenting a simplified diagram of social changes as a basic framework. Second, following the Japanese way of thought, we call the innovation "3C technological innovation" (see B23) and explain why we do so. A detailed framework for "3C technological innovation" is then presented in the form of diagrams according to the basic framework. Key words used in the diagrams are summarized as a thesaurus in the appendix, where reference numbers relating to the key words are also shown.

In Section 3, we classify research on technological innovation into 16 categories according to subjects and methods. The results are illustrated as a table. The diagrams of "3C technological innovation" and the table classifying the types of research can serve as a kind of "map", which can help us traverse a dense forest as we investigate "3C technological innovation".

2 THE FRAMEWORK OF 3C TECHNOLOGICAL INNOVATION

2.1 The Fundamental Structure of Social Changes

There has been much discussion on the character and structure of social changes,e.g., "what are the moving forces behind them?" and "what are their consequences?". While social changes can have many sources, we will be concerned only with those generated by technological innovation. Figure 1 shows a diagram of the simplified, fundamental structure of social changes, where three sub-system form a dynamic cycle in a social, economic, and technical environment. Among these sub-systems, technological innovation seems to play a central role as a moving force or "engine". Technological innovation realizes human desires and in doing so, generates impacts on our society. Schumpeter(see B46) was the first to point out the importance of innovations as the moving force behind social changes, although the importance of technology had already become clear to many economists in the last century. Some of Schumpeter's notions and principles have now been modified, but his belief in the importance of innovations still holds true in today's world, where nations are competing intensely to develop advanced technologies.

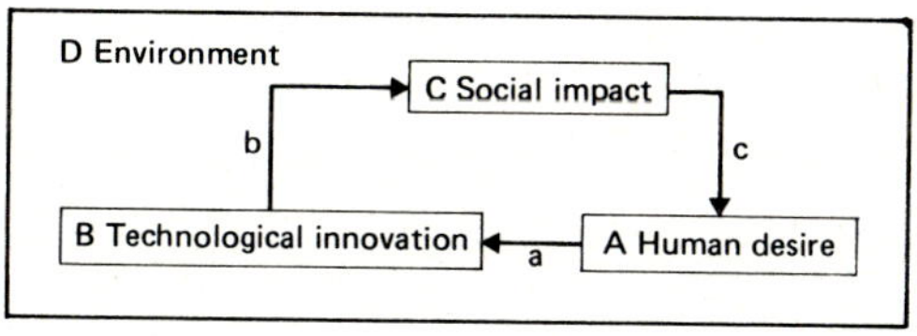

Figure 1
The Fundamental Structure of Social Change

2.2 3C Technological Innovation

The next subject to consider is which major innovation will most affect our society and our lives during the next two decades. We believe that "information innovation" is the most revolutionary kind of innovation and will have the most extensive and profound impacts on society.

Material, energy and information are important components of science and technology, and are essential resources for human society. Technologies exploiting material and energy have played a central role in the process of industrialization, which began in England in the late eighteenth century. Society has just now begun to recognize the importance of information as a resource like material and energy, and the importance of information technology. This can be demonstrated by the following three examples (see B30):

- traffic control with the aid of computers has been found as effective for easing trafffic problems as broadening streets;
- teleconferencing can substitute for travel and save fuel (energy)

- medical information systems can shorten the waiting time of patients in hospitals and can thus be as effecive as increasing the numbers of doctors and/or beds.

Information technology makes it possible to control the production, processing, storage, and use of information in machines and human society by utilizing material and energy. It integrates computer technology, telecommunications technology, and methodologies for forecasting, planning, and management. Historically, information technology has developed as communication technology, control technology, and computer technology, all of which were independent of one another (see B23), or as

Information technology
- Communication technology (technology for communicating information)
- Computer technology (technology for processing, storing, and retrieving information)
- Control technology (technology for maintaining and preserving systems by utilizing information)

Today, largely because of the rapid progress in microelectronics, these technologies have begun to converge, fuse, and invade every human activity. The process by which information technology generates change is called here "3C technological innovation", since we wish to emphasize that such innovation is characterized by the integration of the three C's (communication technology, control technology, and computer technology).

The three most important trends in the integration of the 3C technologies are:

1. Expansion of the use of information resources(data banks, messages, knowledge bases). The goal of this movement is to make it possible to communicate any amount and any kind of information, anywhere and anytime.
2. Improved functional capabilities of machines used in manufacturing and in the service sectors (toward producing goods and services efficiently by using machines with sensing abilities and intelligence).
3. Kowledge information processing (to support human intel-

lectual activities in business, government, education, the home etc.)

The above three directions (1, 2 and 3) might correspond to the well-known acronyms C & C (Computers and Communications), FA (Flexible Automation) and OA (Office Automation), respectively.

The advancement and integration of the 3C technologies are expected to make the following five goals technologically realizable.

- remote and effective communication of large quantities of information
- utilization of many different communication media
- small production of many different products
- improvement of productivity, quality, and reliability(PQR) in factories, offices, and in the service sectors generally
- support of human intellectual activities

In addition, they are expected to have the following effects on the industrial sectors and on society.

- changes in the industrial structure,e.g., the birth of new industries
- decentralization of manufacturing and residential areas (manufacturing network systems, telecommuting, etc.)
- relief for labor from unfavorable work and working environment
- electronization of the home
- global internationalization (international enterprises, weakening of national boundaries,etc.)
- substitution of human and material transport by information transfer through computer-based methods
- energy/resource saving through optimal production processes
- conversion into a service-oriented society (increase in the number of employees in service sectors)
- increase in international competitiveness

A structural diagram of the above-mentioned technological changes and effects is presented in Figure 2.

2.3 Impacts on Society

3C technological innovation will have extensive economic, social, political, and cultural impacts, which will be closely interrelated through information technology. These can be summarized as follows

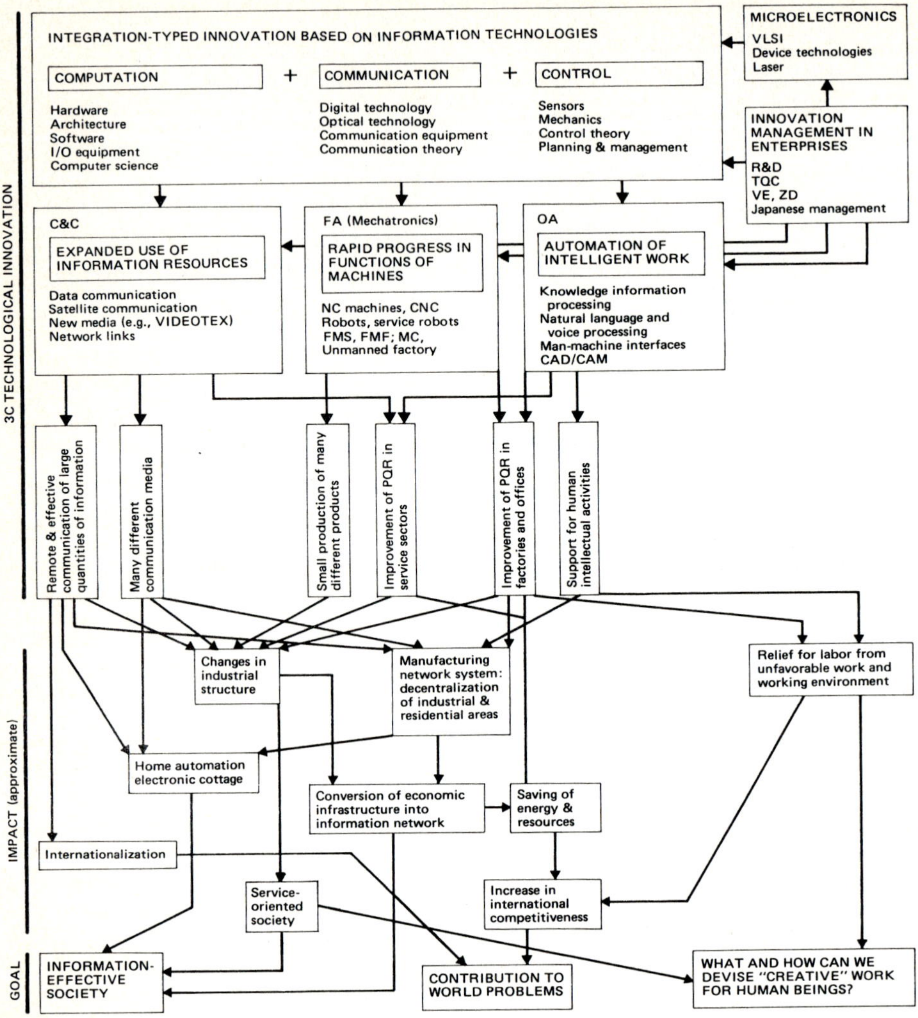

Figure 2
The Structure of 3C Technological Innovation
PQR = productivity, quality, and reliability VE = value evaluation
ZD = zero defect TQC = total quality control

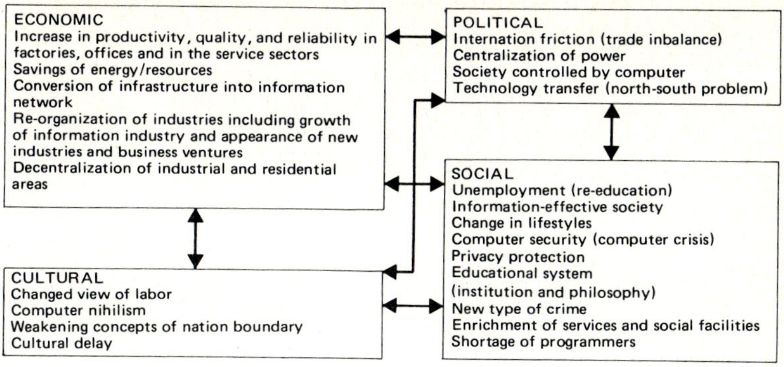

Figure 3
Impacts on Society

(see Figure 3):

A Economic impacts

- increase in productivity, quality, and reliability in factories, offices, and the services sectors
- enegy/resource savings
- conversion of portions of the national infrastructure into information services
- reorganization of industries, including the growth of the information industry and the appearance of new industries and business ventures
- dispersion of industrial and residential areas

B Social impacts

- unemployment (re-education)
- information-effective society
- changes in lifestyles
- privacy protection
- computer security (computer crisis)
- changes in the educational system (institution and philosophy)
- new types of crime
- enrichment of services and social facilities

C Political impacts

- international friction (sharper trade inbalances)
- centralization of power (bureaucratic society)
- computer-controlled society

- technology transfer (the North-South problem)

D Cultural impacts
- changed view of labor
- weakening concepts of national boundary
- computer nihilism
- cultural delay (or runaway technology)

2.4 Human Desires

The structure of technology needs is presented as a diagram in Figure 4. Man's endless search for efficiency, comfort, convenience, variety, self-realization,etc. constantly gives rise to new needs. There will be new consumer demands for new consumer durable goods such as information service equipment, automatic control of air-conditioning and hot water supplies (in the interest of energy/resource savings), and new living systems with educational facilities, handy information services, accident prevention, crime prevention, etc. New social needs related to urban problems, transportation, medicine, welfare, disaster prevention, education and so on also will appear. The following technologies might be among those suitable for meeting these needs:

- solar energy for the home
- electronic cars (for reducing pollution and saving energy)
- residual(waste)-processing systems
- new transportation systems, such as the "demand bus"
- new media for the home
- development of robots serving the older generation and the handicapped
- regional health centers utilizing medical and information equipment
- electronic fund transfer

2.5 Environment

We summarize the environment of 3C technological innovation in Figure 5, where the case of Japan is presented.

The Japanese government is concerned with national innovation policy. In particular, the Ministry of International Trade and Industry (MITI) is responsible for industrial and trade policy, the Ministry of Post and Telecommunications(MPT) for communication policy, the Ministry of Education(ME) for education and science(academic) policy and the Science and Technology Agency(STA) for science and technolgy policy. Various institutional improvements are expected to become necessary with advances in 3C technological innovation so that technological assessments will have to be performed.

Conditions in Japan are likely to be particularly appropriate for 3C technological innovation due to

- the generally high educational level and resulting highly qualified labor force, seen, among other ways, in discussion groups for effective quality control(QC) movements in the enterprises
- the high ability to innovate in order to achieve clearly specified goals, which appears in integration and improvement-oriented innovation
- the existence of an industrial organization that can support the development of advanced technology
- social flexibility in accepting new technologies as a means to improve national wealth and the quality of life
- employment throughout a person's working life
- labor union within the enterprise or the absence of guilds or craft unions
- a reward system based on the number of working years(seniority)

The suitability of Japan seems to be closely related to its national character, as shown in Figure 6, where present-day national character is explained by traditional culture and physical environment.

Summarizing 3C technological innovation, it can be said that our society will move toward a so-called information-effective society as a long range trend in which changes should be directed toward contributing to the solution of existing world problems. In an information-effective society, the productivity of machines in the areas of both goods and services is expected to be extremely high. This will make it necessary for us to find answers to such questions as what we can create as "creative" work for human beings and how we can develop and implement it. These issues are also shown in Figure 2. Finally, it should be noted that 1) national security will depend upon whether 3C technological innovation is successful or not, and 2) there exists a gap between the rates of technological

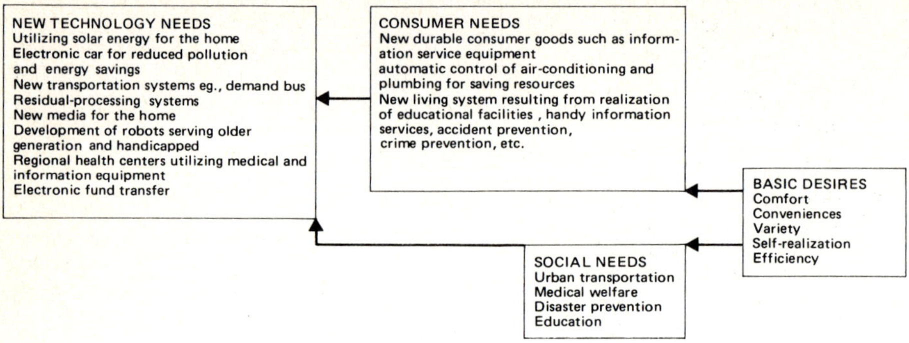

Figure 4
Human Desires

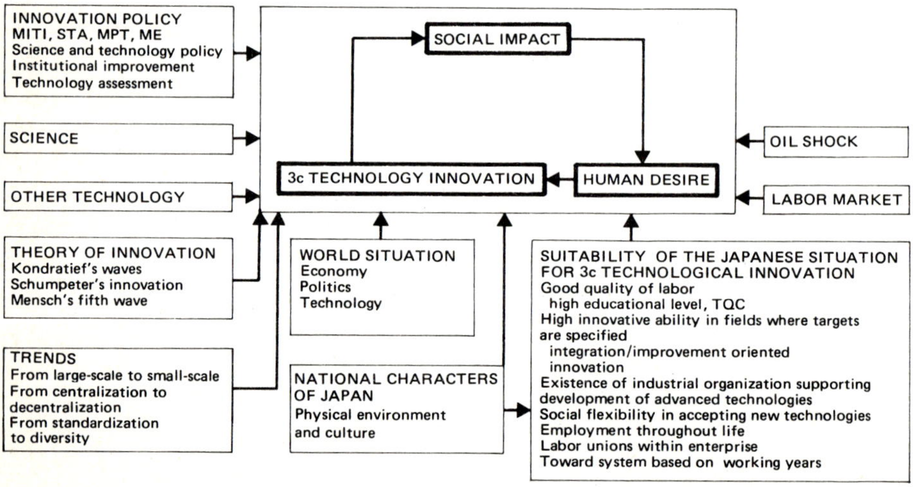

Figure 5
The Environment for 3C Technological Innovation

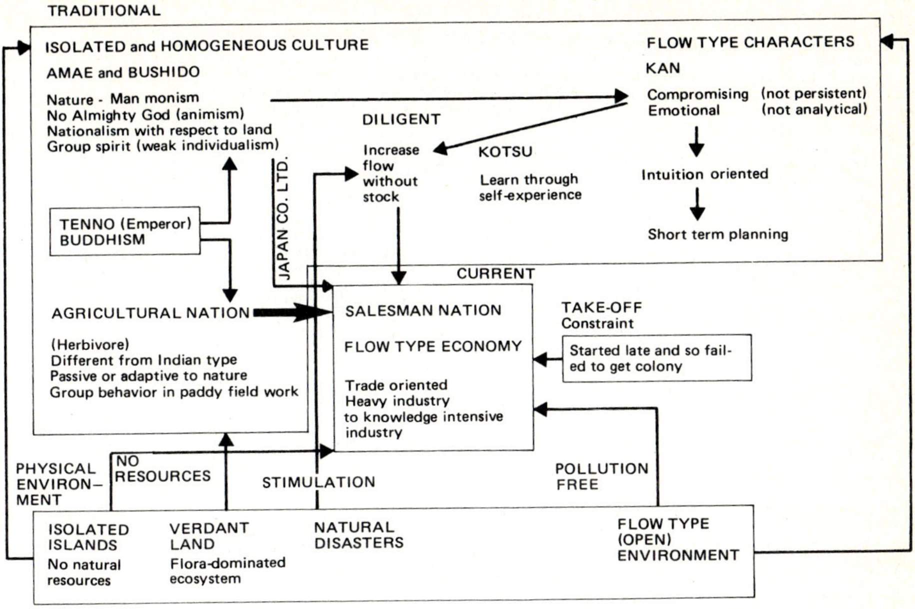

Figure 6
The National Character of Japan[P101]

and cultural changes and thus "cultural delay" or "runaway technology" may become a serious problem.

The thesaurus of 3C technological innovation in the appendix provides us with a useful system of key words for stepping into more detailed research.

3 A FRAMEWORK FOR RESEARCH IN TECHNOLOGICAL INNOVATION

Looking at the diagram of social changes in Figure 1, we can identify three different approaches to research on social change: human desire-oriented, technological innovation-oriented and social impact-oriented. In considering social changes in terms of technological innovation-oriented research, we have illustrated schematically four subjects in Figure 7, where letters A, B, C, D, a, b and

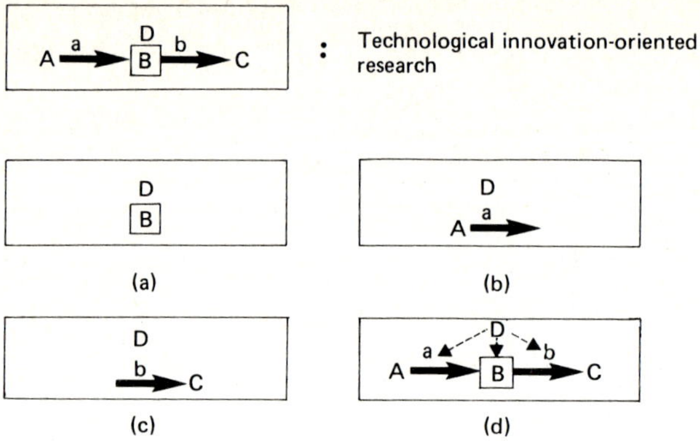

Figure 7
Four Subjects for Technological Innovation-Oriented Research

c correspond to the same letters in Figure 1.

(a) Research on the technological innovation process. The process of technological innovation is shown in the upper part of Figure 8, where we can see the interrelations among the elements along the time axis.

(b) Research into the conditions necessary for technological innovation. In short, the following three conditions must be satisfied for technological innovation to be successful:

1. scientific and technological feasibility
2. the existence of market demands
3. existence of organizations for converting scientific and technological potential into goods and services satisfying the demends

(c) Research on the social impacts of the technological innovation. Economic, social, political, and cultural impacts should be considered.

(d) Research on policy/management of technological innovation. The management of the process represented in the upper part of Figure 8 is a problem encountered at the level of the firm. The conceptual diagram established by the OECD (see R17) to show national management is presented in the lower part of Figure 8.

We also can classify research according to the kinds of methods

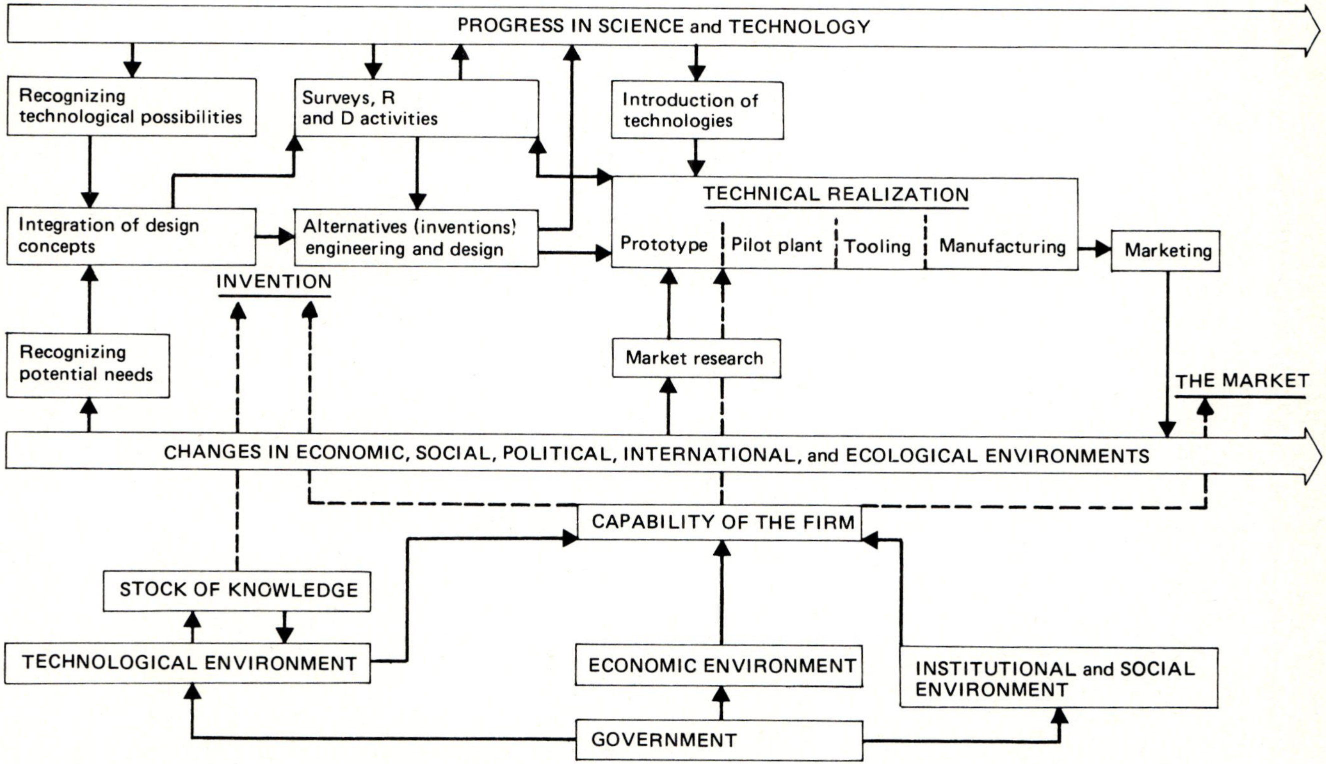

Figure 8
The Process of Technological Innovation and Policy/Management of Technological Innovations

Table 1
Classification of Research

Research Subjects	Empirical research	Mechanism research	Forecasting research	Assessment research
(a) Technological Innovation Process	O	O	O	
(b) Conditions Required for Technological Innovation	O		O	
(c) Social Impact of Technological Innovation	O	O	O	O
(d) Policy/Management for Technological Innovation	O	O	O	O
Comparative Studies among Firms, Sectors, and Nations				

O = valid problem

employed,i.e, empirical, mechanism, forecasting or assessment. Research on technological innovation can be classified into 16 categories as shown in Table 1. It should be noted that the empirical research involves the problem of classifying patterns of technological innovations, for which comparative cross-firm, cross-sectoral, and cross-national studies are indispensable.

APPENDIX: THESAURUS FOR 3C TECHNOLOGICAL INNOVATION

(Numbers in parentheses indicate reference numbers.)

W = international viewpoint
J = national(Japanese) viewpoint
T = technological viewpoint

* 3C technological innovation (B15W/B23J/B30J/P33J)
  * Technological innovations in the 1980s (B1J/B2J/B8W/B38W/P27J/P32J)
  * Innovation management
    * R & D (B27J/B35T/B44J/B51J/P18J/P111J)
    * TQC, VE, ZD (B21J)
    * Japanese management (B41J/B42J)
    * Appropriate technology (P81W)
  * Basic technology
    * Microelectronics (B36W/B37W/P72T/P83W/P103W/P132W)
      * Device technology
        * VLSI (P82T)
          * Josephson device (P36T)
          * GaAs device (HEMT)
      * Microprocessors
      * Laser
      * Material resource situation
    * Computer
      * Hardware
      * Architechture (P49T/P54T/P57T/P85T)
      * Software
        * Software engineering (P39T)
      * Information processing technology (P76T)
      * I/O equipment
      * Computer science (P106T)
      * Information theory
      * Computer language (P36T)
      * Computers in the 1980s (B26W/P11J)
    * Communication (R19J)
      * Digital technology (R26J/P114J)
      * Optical technology (P40T/P43T)
      * Communication equipment
      * Communication theory
      * Future outlook (R25J)
      * Marketing
    * Control
      * Sensors
      * Mechnics
      * Control theory
      * Planning and management
  * Integrated technology
    * C & C (Expanded use of information resources) (R26J/P88T)
      * Data communication
      * Satellite communication
      * New media
        * Videotex (R5T/P51W)
        * CATV
        * Preference analysis (R20J)
      * Personal computer (P30T/P66T/P128T/P133T)
      * Network links
      * Teleconferencing

```
*       FA or mechatronics (progress in functions of machines)(P38J)
*          NC, CNC
*          Robots, service robots (P70J)
*          MC, FMS, FMF
*          Unmanned factory
*          Technology assessment of robots (P48J)
*             CCD
*             Optic fiber
*       OA (knowledge information processing) (P93T/P117W)
*          Natural language processing (P115T)
*          Database, methodbase, knowledgebase
*          Pattern recognition (P2T/P79T)
*          Man-machine interface (R5T)
*          CAD/CAM
*          Decision support system
*    Primary effects
*       Remote, effective communication of large quantities of
         information
*       Many different communication media
*       Small production of many different products (P71W)
*       Improvement of productivity, quality, and reliability in
         factories, offices, and the service sectors (P52W)
*       Support of human intellectual activities
*    Secondary effects
*       Change in industrial structure
*       Home automation (electronic cottage)
*       Conversion of economic infrastructure into information
         network
*       Manufacturing network system
*          Decentralization of manufacturing and residential areas
*       Relief for labor from unfavorable work and working environ-
         ment
*       Resource/energy saving
*       Internationalization
*       Service-oriented society
*       Increase in international competitiveness
*    Goals
*       Information-effective society
*       Contribution to world problems
*       What and how can we devise creative work for human society
*    Enterprises
*       IBM (B4W)
*       Fujitsu (B28J/B53J)
*       Hitachi
*       NEC (B24J)


* Impacts on society (B14J)
*    Economic impacts
*       Information and economic growth (R18W/R21J/P89J)
*       Increase in productivity, quality, and reliability in
         factories, offices, and the service sector
*       Resource/energy saving
*       Conversion of national infrastructure into information
         network
*          Distribution of information stocks (R27J)
*          Information flows (R21J)
*          Information economics
*          Electronic fund transfer
*       Reorganization of industry
*          Growth of information industries
```

```
*          Appearance of new industries (R16J)
*          Appearance of new business ventures
*          Machlup's knowledge industry (B45W)
*       Decentralization of manufacturing and residential areas (R6J)
*          Teleconferencing
*          Telecommuting
*    Social impacts
*       Unemployment (reeducation) (R18W/B15W/P16W/P17W/P28W/P135W)
*       Information-effective society (R26J/P45J/P97W/P98W)
*          Industrial society (P31J)
*          Informational indices (R21J)
*       Changes in lifestyles
*       Computer security (computer crisis) (P27W/P130T)
*       Changes in educational system
*          CAI
*       New types of crime
*       Enrichment of services and social facilities
*       Social development (R21J)
*    Political impacts
*       International friction
*          Trade imbalances (R18W/P116W)
*       Centralization of power
*       Computer-controlled society
*          Computer democracy (P3T/P20T/P48T/P89T/P105T)
*       Technology transfer (P122W/P135W)
*          North-South problem
*    Cultural impacts
*       Changed view of labor
*       Weakening concepts of national boundary
*       Computer nililism (P73J)
*    General assessment of social impacts (R2J)
*    T. Parsons


* Human desire
*    Basic desires
*       Comfort
*       Convenience
*       Variety
*       Self-realization
*       Mazroh
*    Consumer needs
*       New durable consumer goods
*          Information service equipment
*       Energy/resource savings
*          Automatic control of air conditioning and water supply
*       New living system
*          Education facilities
*          Handy information services for reservations, orders,
            payments, etc.
*          Accident prevention
*          Crime prevention
*    Social needs
*       Urban
*       Transportation
*       Medical
*       Welfare
*       Disaster prevention
*       Education
*    New technology needs (R28J)
```

```
*        Solar energy for the home
*        Electronic cars (for reducing pollution and saving energy)
*        New transportation systems, such as "demand bus"
*        Residual-processing systems
*        New media for the home
*           Videotex
*        Robots serving the older generation and the handicapped
*        Regional health centers utilizing medical and information
*         equipment
*        Electronic fund transfer


* Environment
*     Innovation policy (B43W/P9W/P28W/P34J/P123J)
*        MITI
*           Industrial innovation policy (R17W/B3J)
*           Research association for VLSI development
*           Development of fifth generation computer (R14J/P62J)
*           Trade policy
*           White paper (B29J)
*        MPT
*           Telecommunication policy
*           NTT
*           KDD (Japan Overseas Radio and Cable System)
*           NHK (Japan Broadcasting Corporation)
*           Regulation (P65W)
*           White paper (B33J)
*        ME
*           Education and science policy
*        STA
*           Science and technology policy (B49J/P1J/P44J/P86W/P112J/
*           White paper (B48J/B50J)                               P113W)
*        Institutional improvement
*        Big R & D management (R4J)
*        Technology assessment (R3W)
*        Policy making process (B34J)
*        Independent technology development (P60J)
*           Investment for R & D (P29W/P82W)
*     Science
*        Basic research (P23J/P59J/P118J)
*        Science and technology in the 1980s (P45W)
*        History of science and technology (B6J/B52T/B55W)
*     Other technology
*        Bioengineering
*        Solar energy (P63T)
*     Theory of innovation
*        Kondratief's waves
*        Schumpeter's innovation (B46W)
*        Mensch's fifth wave
*        Toffler's third wave
*     Trends
*        Large scale to small scale
*        Centralization to decentralization
*        Standardization to diversity
*     World situation
*        Economy (P47W)
*        Politics
*        Technology
*           Energy
*           Resources
```

```
*         Crisis (P68W)
*      Japanese national character (P101J)
*         Physical environment
*            Isolated islands
*            No natural resources
*            Verdent land
*            Natural disasters
*            Flow type environment (rainy and windy)
*         Isolated and homogeneous culture
*            Nature-man monism
*               No almighty god (animism)
*            Nationalism with respect to land
*            Group spirit (weak individualism)
*         Agricultural nation
*            Passive or adaptive to nature
*            Group behavior in paddy field work
*         Flow type character
*            Compromising (not persistent)
*            Emotional (not analytical)
*         Saleman nation
*            Flow type economy
*               Trade oriented
*               Heavy industry to knowledge-intensive industry
*      Suitability of Japanese situation for 3C technological innavation
*         High quality labor
*            TQC (QC circles)
*         High innovative ability in fields where goals are specified
*            Integration/improvement-oriented innovation
*         Support for development of advanced technologies from
           industrial organizations
*         Social flexibility in accepting new technology
*            Accepting new technologies as means
*         Labor union in enterprise (no guilds)
*         Reward system according to working years (seniority)
*      Labor market (P84W)
```

REFERENCES

() = international viewpoint
[] = national(Japanese) viewpoint
<> = technological viewpoint

Reports:

<R1> Engineering Development Association: Investigation of project management(in Japanese). July 1979. 285 pages.

[R2] Institute for Future Technology: General assessment of advancement of informatization of society(in Japanese). March 1974. 212 pages.

(R3) Institute for Future Technology: Investigation of technology assessments in foreign countries(in Japanese). March 1980. 65 pages.

[R4] Institute for Future Technology: Effective management of large-scale R & D(in Japanese). May 1980. 300 pages.

<R5> Institute for Future Technology: Research on forecasting of telecommunication technologies - Social acceptability and man-machine interface(in Japanese). March 1981. 206 pages.

[R6] Institute for Future Technology: Research on roles of telecommunication technology in regional relocation of urban facilities - Regional relocation of administrative functions and telecommunications(in Japanese). March 1981. 219 pages.

(R7) International Data Corporation: Automated business communications - Issues in implementing office automation. August 1981. 79 pages.

(R8) International Data Corporation: Corporate, planning service - Industry overview and outlook. August 1981. 71 pages.

(R9) Japan Electronic Computer Corporation: Investigations of future technology, market and environment of computers(in Japanese). February 1979. 310 pages.

(R10) Japan Electronic Industry Development Association: Basic investigations to develop a vision for cooperation with foreign countries in the field of informatization (I)(in Japanese). March 1980.327 pages.

[R11] Japan Electronic Industry Development Association: Long-range view of electronic industries(in Japanese). March1980. 559 pages.

<R12> Japan Electronic Industry Development Association: Trends in computer system technologies - Summary of the 1970s and a view toward the 1980s(in Japanese). March 1980. 255 pages.

[R13] Japan Industrial Robot Association: Technology assessment of industrial robots(in Japanese). May 1980. 319 pages.

[R14] Japan Information Processing Development Center: Investigations of fifth generation computers(in Japanese). March 1980 and March 1981.

[R15] Japan Science Foundation: Development of methods for analysis of relational structure among elementary technologies (summary)(in Japanese). February 1979. 72 pages.

[R16] Japan Techno-Economics Society: Survey reprts on formation of new industries due to innovative technologies(in Japanese). February 1980. 198 pages.

(R17) OECD: Policies for the stimulation of industrial innovation. Vols.I,II-1, and II-2. 1978.

(R18) OECD: Information activities, electronics and telecommunications technologies - Impacts on employment, growth and trade . Vol.I. September 1980. 138 pages; Group of experts on economic analysis of information activities and the role of electronics and telecommunications technologies. Vol.II.June 1980. 533 pages.

[R19] Research Institute of Telecommunications and Economics: Roles of electric telecommunication in future society(in Japanese). March 1977. 655 pages.

[R20] Research Institute of Telecommunications and Economics: Analyses of preferences in communication media(in Japanese). March 1978. 320 pages.

[R21] Research Institute of Telecommunications and Economics: Measurements of "information flow" in terms of major information indices and economic impacts due to informatization (in Japanese). March 1979. 185 pages.

(R22) Research Institute of Telecommunications and Economics: Research on systems approaches for the market (I). March 1979. 65 pages. (II). March 1980. 70 pages. (in Japanese)

[R23] Research Institute of Telecommunications and Economics: Informatization of economic activities and tlelcommunication (in Japanese). March 1980. 227 Pages.

(R24) Research Institute of Telecommunications and Economics: Main issues in telecommunication policies in the U.S. and European countries(in Japanese). March 1980. 252 pages.

[R25] Research Institute of Telecommunications and Economics: Research on future views of industries relating to tlelcommunication(in Japanese). March 1980. 215 pages.

[R26] Research Institute of Telecommunications and Economics: Theoretical research on future telecommunication systems(in Japanese). March 1980. 263 pages.

[R27] Research Institute of Telecommunications and Economics: Research on distribution of information stock - Home and information services(in Japanese). March 1980. 246 pages.; Research on distribution of information stock(in Japanese). March 1981. 333 pages.

[R28] Research Institute of Telecommunications and Economics: Behavioral-scientific analysis and long-term forecasting of needs for information and telecommunication(in Japanese). March 1981. 391 pages.

(R29) Research Institute of Telecommunications and Economics: Trends in data communication in the U.S. and European countries(in Japanese). March 1981. 389 pages.

(R30) Research Institute of Telecommunications and Economics: Trends in telegraph and telephone in the U.S. and European countries(in Japanese). March 1981. 418 pages.

Books:

[B1] Agency of Industrial Science and Technology(ed.): Future plan of technological development(in Japanese). Tsusho-sangyo-kai. September 1977. 235 pages.

[B2] Agency of Industrial Science and Technology(ed.):Technological development - Groping through the 1980s. Nikkan-kogyo-shinbun-sha. December 1979. 192 pages.

[B3] Agency of Industrial Science and Technology(ed.): Toward establishment of a state on the basis of creative technologies(in Japanese). Tsusho-sangyo-chosa-kai. February 1981. 141 pages.

(B4) Aono,T.: Light and darkness of IBM(in Japanese). Nippon-keizai-shinbun-sha. October 1972. 238 pages.

(B5) Bell,D.: The coming of post-industrial society - A venture in social forecasting. Basic Books. 1973. 507 pages.

[B6] Cabinet secretariats concerns(ed.): Historical view of science and technology(in Japanese). Published by Min. of Finance. August 1980. 113 pages.

[B7] Defence Agency(ed.): White paper on defence(in Japanese). Published by Min. of Finance. September 1981. 323 pages.

(B8) Dertouzos,M.L. et al.: The computer age: A twenty-year view. The MIT Press. December 1979. 491 pages.

[B9] Economic Planning Agency(ed.): New economic and social 7-year planning(in Japanese). August 1979. 181 pages.

[B10] Economic Planning Agency(ed.): The present state of the Japanese economy(in Japanese). Published by Min. of Finance. March 1981. 356 pages.

[B11] Economic Planning Agency(ed.): White paper on economy(in Japanese). Published by Min. of Finance. September 1981. 750 pages.

[B12] Hayashi,Y.: Information-effective society(in Japanese). Kodan-sha-shinsho. May 1969.209 pages.

<B13> Holt,K.: Innovation and creativity(translation in Japanese). Kaihatsu-sha. March 1977. 263 pages.

[B14] Hoshino,Y.: Technological innovation(in Japanese). Iwanami-shinsho. April 1975. 220 pages.

(B15) ILO: The impact of micro-electronics: A tentative appraisal of information technology. 1980.

(B16) Inose,H. et al.: Do microcomputers produce unemployment?(in Japanese). Computer-Age-sha. July 1981. 250 pages.

[B17] Iwasaki,T.: Front line of robot industry(in Japanese).Shijo-shinbun-sha. October 1981. 280 pages.

[B18] Japan Electronics Industry Development Association(ed.): Electronics in Japan - A step toward systematizing(in Japanese). Computer-Age-sha. May 1978. 325 pages.

[B19] Japan Information Processing Development Center: Proceedings of the International Conference on Fifth Generation Computer Systems. Tokyo. October 1981.

[B20] Japan Techno-Economics Society: Proceedings of the 10th Technology Assessment Symposium(in Japanese). Tokyo. June 1981. 49 Pages.

[B21] Karatsu, H.: TQC - Japanese intelligence. Nikkagiren-shuppan -sha. July 1981. 183 pages.

[B22] Kimizuka,Y.: Computer business world. Kyoiku-shinsho(in Japanese). 1977. 270pages.

[B23] Kitagawa,T. et al.(ed.): The innovation of information technology(in Japanese). Social Information Science Series, No. 2. Gakken. 1972.

[B24] Kobayashi,K.: C & C is Japanese intelligence - Development of a way toward the 21st century(in Japanese). SIMAL-shuppan -kai. November 1980. 246 pages.

[B25] Komatsuzaki, S.: Information industries - Seeking a strategy for industries in the 1980s(in Japanese). Toyo-keizai-shinpo -sha. November 1980. 214 pages.

(B26) Lecht,C.P.: The waves of change - A techno-economic analysis of the data processing industry(translation in Japanese). Kikaku-center. December 1978. 217 pages.

[B27] Makino,N.: Technological development for Japanese enterprises-challenges from foreign technologies(in Japanese). Nikkei-shinsho. April 1968. 205 pages.

[B28] Matsuo,H.: Story of Fujitsu - Courageous changes(in Japanese). Aoba-shuppan. 1978. 221 pages.

[B29] Ministry of International Trade and Industry(ed.): Mini-white paper on international trade and industry(in Japanese) . Tsusan-chosa-kai. July 1981. 222 pages.

[B30] Ministry of International Trade and Industry(ed.): Informatization of society and information industries in the 1980s -A guideline for afluent information effecive society: A report of an inquiry of the committee of information industry(in Japanese). Industrial Rationalization Council. Computer-age-sha. August 1981. 368 pages.

<B31> Ministry of International Trade and Industry(ed.): Fundamentals of electronic technologies - Guidance for the elecronics age(in Japanese). Published by Min. of Finance. September 1981. 308 pages.

[B32] Ministry of Labor(ed.): White paper on labor(in Japaneae). Published by Min. of Finance. July 1981. 645 pages.

[B33] Ministry of Post and Telecommunications(ed.): White paper on telecommunication(in Japanese). Published by Min. of Finance December 1980. 549 pages.

[B34] Murakawa,I.: Policy making process(in Japanese). Kyoiku-sha. 1979. 237 pages.

<B35> Nikkan-kogyo-Shinbun-sha(ed.): Challenges against a wall in technological development(in Japanese). November 1970. 199 pages.

(B36) Nikkan-Electronics(ed.): 50 year history of electronics and future view toward the 21st century. Nikkei-McGraw-Hill, November 1980. 542 pages.

(B37) Nikkei-Electronics(ed.): Electronics innovations(inJapanese). Nikkei-McGraw-Hill. April 1981. 330 pages.

(B38) Nippon-kezai-shinbun-sha(ed.): Searching for the most advanced technologies in the world - Technological innovation in the 1980s(in Japanese). Nippon-keizai-shinbun-sha. 1980. 233 pages.

(B39) OECD: The conditions for success in technological innovation(translation in Japanese). Tsusho-sangyo-chosa-kai. November 1974. 279 pages.

(B40) Orme,M.: Micros; A pervasive force(translation in Japanese). Nippon-noritsu-kyokai. October 1981. 258 pages.

[B41] Ouchi,W.G.: Theory Z - How American business can meet the Japanese challenge(translation in Japanese). CBS-SONY-shuppan. August 1981. 357 pages.

[B42] Pascale,R.T. and Athos,A.G.: The art of Japanese management (translation in Japanese). Kodan-sha. July 1981. 311 pages.

(B43) Pavitt,K. and Walker,W.: Innovation policies. Sangyo-chosa-kai. 1978. 164pages. Translation in Japanese from Government policies towards industrial innovation: A review. Research Policy. Vol.5. 1976. Pp.11-97.

[B44] Research Group on R & D(ed.): Strategies for R & D(in Japanese). Kaihatsu-sha. December 1973. 146 pages.

(B45) Research Institute of Telecommunications and Economics: Machlup - A concept of knowledge industry(in Japanese). Diamond-sha. February 1971. 188 pages.

(B46) Schumpeter,J.A.: Theory of the economic development(translation in Japanese). Iwanami-bunko. 1977.

<B47> Science and Technology Agency: Present situation and future view of telecommunication(in Japanese). Published by Min. Finance. 1977. 282 pages.

[B48] Science and Technology Agency: White paper on science and technology - History and future view of development of science and technology(in Japanese). Published by Min. Finance. 1977. 282 pages.

[B49] Science Council of Japan(ed.): Toward establishment of forecasting science and technology policy. Published by Min. Finance. March 1981. 251 pages.

[B50] Science and Technology Agency: White paper on science and technology - International comparisons and future issues (in Japanese). Published by Min. of Finance. July 1981. 397 pages.

[B51] Tadano,B. et al.: R & D - Challenges towards problems existing in the Japanese enterprises(in Japanese). Management Center. April 1971. 274 pages.

[B52] Tsubota,I.(ed.): Science and technology in the last 50 years - Series of history of Japan in Showa(in Japanese). Vol.15. April 1976. 177 pages.

[B53] Usui,K.: Fujitsu changes trends(in Japanese). Recruit-shinsho. July 1981. 250 pages.

[B54] Yamamoto,H.: The spirit of Japanese capitalism - Why do the Japanese work so hard?(in Japanese). Kobun-sha. November 1979. 230 pages.

(B55) Yamazaki,T. et al.: Outline of the history of science and technology(in Japanese). Ohmu-sha. June 1978. 250 pages.

[B56] Warfield,J.N.: Societal systems: Planning, policy, and complexity. Wiley. 1976.

Papers:

[P1] A dicussion meeting: Science policy in Japan. Technology and Economy, June 1979. Pp.32-61(in Japanese).

<P2> Agency of Industrial Science and Technology: Pattern information processing system (I), Denshi-kogyo-geppo, Vol.21, No.5, 1979. Pp.5-11; (II),Vol.22,No.5,1980. Pp.21-31(in Japanese).

<P3> Ando,F.: Decision room system, Technology and Economy, September 1980. Pp.95-109(in Japanese).

<P4> A special article: Problems in software engineering, COMPUTOPIA, December 1979. Pp.12-37(in Japanese).

<P5> A special article: Changing office computers, COMPUTOPIA, May 1980. Pp.18-73(in Japanese).

[P6] A talk: Effects of the technological progress of LSI on telecommunications - Trends in the progress of LSI technology and the information processing technology, COMPUTOPIA, October 1979. Pp.10-19(in Japanese).

(P7) Ayers,F.T.: The management of technological risk, Research Management, Vol.20, No.4, September 1979. Pp.493-513.

(P8) Barna,B.: The ranking of computer industries - Best 50, COMPUTOPIA, October 1979. Pp.32-61(translation in Japanese).

(P9) Beckler,D.: Innovation under the new social and economic order. Presented at the 1980 International Conference on Innovation Policy held in Tokyo, Technology and Economy, May 1980. Pp.120-133(translation in Japanese).

(P10) Boretsky,M.: Trends in U.S. technology: A political economist's view, American Sceintist, Vol.63, January-February 1975. Pp.70-82.

[P11] COMPUTOPIA: How will the Japanese computer industries become ? - Views and expectations for the technology market in the 1980s, January 1979. Pp.62-93(in Japanese).

(P12) DATAMATION Special Talk: The LSI revolution, COMPUTOPIA, October 1979. Pp.22-34(in Japanese).

(P13) Dohrmann,R.J.: Matching company R & D expenditures to technology needs, Research Management, November 1978. Pp.17-21.

<P14> Editors of COMPUTOPIA: Advancing of functions and systems of FAXIMILI, COMPUTOPIA, February 1980. Pp.12-40(in Japanese).

<P15> Editors of COMPUTOPIA: The advent of the "voice" information processing age, COMPUTOPIA, August 1980. Pp.12-27(in Japanese).

(P16) Freedrich,G.: New phases of technological innovation and employment in microelectronics, Technology and Economy, December 1979. Pp.4-29(translation in Japanese).

(P17) Girichrist,B.: Impacts of computers on employment, Technology and Economy, December 1979. Pp.30-40(translation in Japanese).

[P18] Goto,A.: Technological progrss and R & D in Japan, Kikan-gendai-keizai, Summer 1980. Pp.153-163(in Japanese).

(P19) Gould,P. and Johnson,J.: National television policy monitoring structural complexity, Futures, Vol.12, No.3, June 1980. Pp.178-190.

<P20> Harashina,Y.: Conference supporting system for environmental planning, Technology and Economy, September 1980. Pp.40-60 (in Japanese).

(P21) Harmann,A.J.: Industrial innovation and governmental policy: A review and proposal based on observation of the US electronics sectors, Techn.Forec.Soc.Change, Vol.18, 1980. Pp. 15-37.

(P22) Harmann,A.J.: Innovation management: Research protocol for international collaboration, IIASA Working Paper, November 1980, 46 pages.

[P23] Hasegawa,H.: A myth of basic research. Technology and Economy, October 1980. Pp.40-58(in Japanese).

(P24) Haustein,H.D.: Flexible automation: The issues, IIASA, 1981. 11 pages.

(P25) Haustein,H.D. and Maier,H.: Basic improvement and pseudo-innovations and their impact on efficiency, Techn.Forec.Soc. Change, Vol.16, 1980, Pp.243-265.

(P26) Healey,F.H.: Industry's needs for basic research, Research Management, November 1978. Pp.12-16.

(P27) Hirabayashi,C.: Fearful computer crises - Their likelihood and counter-actions, CONPUTOPIA, March 1981. Pp.34-44(in Japanese).

(P28) Honey,N.B.: Relations in technological development between industries and government: An American's view, Technology and Economics, May 1980. Pp.133-138(translation in Japanese).

(P29) Iinuma,M.: Trends in investments for R & D in the U.S. and Europe, Technology and Economy, October 1980. Pp.81-105(in Japanese).

<P30> Ikeda,K.: Technology and applications of super minicomputers, Denshi-kogyo-Geppo, Vol.21, No.8, 1979. Pp.40-53(in Japanese).

[P31] Imai,K.: Industrial society of Japan - A trial toward '80s, Kikan-gendai-keizai, Winter 1979. Pp.44-61(in Japanese).

[P32] Imai,K.: Breakthrough in "technology economy", Gekkan-NIRA, April 1980. Pp.4-9(in Japanese).

[P33] Imai,K.: The establishment of a state on the basis of technologies in Japan, Economist, April 10, 1981. Pp.8-16(in Japanese).

[P34] Inose,H.: Some topics on technological innovation policy relating to the trend in Japan. Presented at the 1980 International Conference on Innovation Policy held in Tokyo. Technology and Economy. June 1980. Pp.128-136(translation in Japanese).

(P35) International Chamber of Commerce: Technology transfer for development, Technology and Economy, February 1980. Pp.110-136(translation in Japanese).

<P36> Ishida,H.: BASIC for microcomputers, Denshi-kogyo-geppo, Vol.21, No.11, 1979. Pp.28-46(in Japanese).

<P37> Ishii,O.: How will the advancement of VLSI influence computers in the 1980s?, COMPUTOPIA, March 1979. Pp.18-24(in Japanese).

[P38] Ishii,S.: The electronics industry and the machinery industry, Denshi-kogyo-geppo, Vol.21, No.5, 1979. Pp.25-39(in Japanese).

<P39> Ito,K.: Software engineering - Supporting tools on trial, Denshi-kogyo-geppo, Vol.21, No.8, 1979. Pp.21-31(in Japanese).

<P40> Ito,R.: Optical integrated circuits, Denshi-kogyo-geppo, Vol.22, No.1, 1980. Pp.29-44(in Japanese).

(P41) Jussawlla,M.: Communication technology transfer - Impact on economic development, Telecommunications Policy, December 1980. Pp.249-262.

(P42) Kamien,M.I. and Schwartz,N.L.: Market structure and innovation: A survey, J.Economic Literature, March 1975. Pp.1-37.

<P43> Kamiya,T.: Trends in the R & D of optical devices, Denshi-kogyo-geppo, Vol.22, No.2, 1980. Pp.67-72(in Japanese).

[P44] Kawase,T.: Future problems expected in science and technology budgets, Technology and Economy, JUne 1979. Pp.4-31(in Japanese).

(P45) King,A.: The roles of science and technology in the 1980s, Technology and Economy, January 1980. Pp.11-17(translation in Japanese).

(P46) Kitagawa,T.: A theoretical formulation of information network systems and its implication to social intelligence, Res. Rep. No.1, IIAS-SIS, Fujitsu Ltd., 1978.

(P47) Klein,L.R.: World economy - Its scenario in the 80s, Kikan-gendai-keizai, Spring 1981. Pp.58-71(translation in Japanese).

<P48> Kobayashi,S.: Computer aid for collective decision making, Technology and Economy, September 1980. Pp.63-84(in Japanese).

<P49> Kobayashi,T.: Trends in data-flow machines and parallel processors in the U.S. and Europe(1), Denshi-kogyo-geppo, Vol. 21, No.12, 1979. Pp.48-55(in Japanese).

<P50> Kobayashi,T.: Trends in data-flow machines and parallel processors in the U.S. and Europe(2), Denshi-kogyo-geppo, Vol.22, No.1, 1980. Pp.58-67(in Japanese).

(P51) Komatsuzaki,S.: The impact on communication theory of the evolution of media, Int. Soc. Science J., Vol.33, No.1, 1981. Pp.91-98.

(P52) Kuroda,M.: Productivity sequence and market performance in Japanese economy - A U.S.-Japan comparison of productivity sequences, Kikan-gendai-keizai, Summer 1981. Pp.56-72(in Japanese).

(P53) Little,B. and Cooper,R.G.: The role of marketing research in new technology ventures, Research Management, Vol.20, No.3, May 1977. Pp.20-25.

<P54> Maeda,K.: Computer architecture required in the 1980s - Images of non-Neumann type computers, COMPUTOPIA, January 1980. Pp.44-51(in Japanese).

(P55) Maier,H. and Haustein,H.D.: Innovation, efficiency cycle, and strategy implications, Techn. Forec. Soc. Change, Vol. 17, 1980. Pp.35-49.

(P56) Mansfield,E. and Rapoport,J.: The costs of industrial product innovations, Management Science, Vol.21, No.12, August 1975. Pp.1380-1386.

<P57> Marbin,A.: Concepts of data-flow architecture, Denshi-kogyo-geppo, Vol.22, No.1, 1980. Pp.13-28(translation in Japanese).

(P58) Markowich,P.: A mathematical model for the diffusion of innovation, IEEE Trans. SMC, Vol.SMC-11, No.7, July 1981. Pp.505-509.

[P59] Marumo,K.: From the "organization" to the "individual", Technology and Economy, October 1980. Pp.59-80(in Japanese).

[P60] Miyagawa,K.: What should large-scale and independent development aim at?, Technology and Economy, July 1979. Pp.94-120 (in Japanese).

[P61] Mori,H.: Strategies for R & D in the 1980s, J. Japan Soc. Electric Engineering, Vol.101, No.10, 1981. Pp.5-9(in Japanese).

[P62] Moto-oka,T.: Explanations of developments in knowledge information systems - From TAC to fifth generation computers, COMPUTOPIA, February 1981. Pp.91-95(in Japanese).

<P63> Mukaibo,T. et al.: A talk: The future of energy development, Kikan-gendai-keizai, Spring 1981, Pp.194-205(in Japanese).

<P64> Nagao,S.: The fourth international conference on pattern recognition, Denshi-kogyo-geppo, Vol.21, No.1, 1979. Pp.49-52(in Japanese).

(P65) Nanbu,T.: A U.S.-Japan comparison of regulated industries - Laying stress on telecommunication, Kikan-gendai-keizai, Summer 1981. Pp.87-99(in Japanese).

<P66> Narita,S.: Microcomputer and computer control, Denshi-kogyo-geppo, Vol.21, No.9, 1979. Pp.19-50(in Japanese).

(P67) Nelson,R.R.: In search of a useful theory of innovation, Research Policy, Vol.6, 1979. Pp.35-76.

(P68) Nikkei-Business: Scenario of crises in '81 - Security of management, January 12, 1981. Pp.40-43(in Japanese).

[P69] Nikkei-Bisiness: Scenario of crises in '81 - "No profit without risks": Have advanced industries counter-plans for them?, January 12, 1981. Pp.44-48(in Japanese).

[P70] Nikkei-Busuness: The "age of robots" will come in the 1980s, January 12, 1981. Pp.120-121(in Japanese).

(P71) Nikkei-Business: Small batch production realizing industrial rationalization, February 9, 1981. Pp.103-107(in Japanese).

<P72> Nikkei-Business: A forecast: Electronics will grow rapidly, April 20, 1981. Pp.136-138(in Japanese).

[P73] Nikkei-Business: Information system divisions are falling ill, April 20, 1981. Pp.168-171(in Japanese).

<P74> Nikkei-Business: A forecast: The substitution effect of solar energy will reach 10% or 10^{12} yen, May 4, 1981. Pp.131-133(in Japanese).

[P75] Nikkei-Business: A survey: Ranking of incomes in information industries, Summer 1981. Pp.49-53(in Japanese).

<P76> Nippon-denshi-kogyo-kyokai: A future view of information processing technology - A 10 year vision of computer systems, Denshi-kogyo-geppo, Vol.21, No.8, 1979. Pp.5-20(in Japanese).

<P77> Nippon-denshi-kogyo-kyokai: Trend in applications of licences relating to the Josephson device (1), Denshi-kogyo-geppo, Vol.21, No.6, 1979. Pp.35-40; (2), Vol.21, No.7,1979. Pp.31-39(in Japanese).

[P78] Nishimura,Y.: The future of the microcomputer society, Technology and Economy, January 1981. Pp.75-82(in Japanese).

<P79> Nishino,H.: How far will the pattern information processing technology be developed in the 1980s, COMPUTOPIA, March 1979. Pp.46-54(in Japanese).

<P80> Niwa,F.: Consensus making system by ORAKEL, Technology and Economy, September 1980. Pp.19-39(in Japanese).

(P81) Norman,C.: New ways to technological innovations by appropriate technologies (I), Technology and Economy, July 1979. Pp.118-137(translation in Japanese).

(P82) Norman,C.: Knowledge and power: Funds for R & D in the world, February 1981. Pp.4-78(translation in Japanese).

(P83) Norman,C.: Microelectronics in the world economy (I), Technology and Economy, March 1981. Pp.116-139; (II), April 1981. Pp.91-116; (III), May 1981. Pp.102-113; (IV), June 1981. Pp.110-136(translation in Japanese).

(P84) Oohashi,I.: A U.S.-Japan comparison of labor markets, Kikan-gendai-keizai, Summer 1981. Pp.100-113(in Japanese).

(P85) Orgenick,E.I.: Trends in computer system architecture technology in the U.S., Denshi-kogyo-geppo, Vol.21, No.2, 1979. Pp.12-22(translation in Japanese).

(P86) Organization of Technology Assessment, U.S. Congress: Trends in science and technology policies in foreign countries, Technology and Economy, July 1979. Pp.28-39(translation in Japanese).

(P87) Organization of Technology Assessment, U.S. Congress: Impacts of microcomputers on the society, Technology and Economy, December 1979. Pp.47-64(translation in Japanese).

(P88) Osatake,T. et al.: A special edition: New stage of information network, COMPUTOPIA, April 1980. Pp.16-62(in Japanese).

<P89> Ozaki,I.: Answers of economics to advancing structural changes - The technological system and the economic system, Kikan-gendai-keizai, Autumn 1980. Pp.76-89(in Japanese).

(P90) Packard,D.: Challenges toward the 1980s - NCC '80 keynote speech, COMPUTOPIA, August 1980. Pp.28-53(translation in Japanese).

(P91) Panel discussions at the 1980 International Conference on Innovation Policy held in Tokyo, Technology and Economy, July 1980. Pp.78-106(translation in Japanese).

(P92) Pavitt,K.: Government policy towards industrial innovation: A review, Research Policy, Vol.5, 1976. Pp.11-97.

<P93> Poppel,H.L.: Should office automation be introduced or not?, COMPUTOPIA, January 1980. Pp.64-73(translation in Japanese).

(P94) Press,S.J.: Qualitative controlled feedback for forming group judgements and making decisions, J. Amer. Statist. Assoc., September 1978. Pp.526-535.

(P95) Roberts,E.B.: What do we really know about managing R & D?, Research Management, November 1976. Pp.7-11.

(P96) Robinson,J.M.: Technological learning, technological substitution, and technological change, Techn. Forec. Soc. Change, Vol.18, 1980. Pp.39-49.

(P97) Rubin,M.R. et al.: The U.S. information sector and GNP: An input-output study, Information Processing & Management, Vol.17, 1981. Pp.163-194.

(P98) Rubin,M.R. et al.: Selected roles of information goods and services in the U.S. national economy, Information Processing & Management, Vol.17,1981. Pp.195-213.

(P99) Sakamura,K.: NCC '79 and the USA, Denshi-kogyo-geppo, Vol. 21, No.7, 1979. Pp.63-68(in Japanese).

[P100] Shimazu,Y. and Sugiyama,K.: Limit to growth of Japan in a viewpoint of environmental pollution, Keizai-hyoron, September 1973. Pp.6-22(in Japanese).

[P101] Shimazu,Y., Urabe,T. and Sugiyama,K.: An ecological approach to national planning: Conversion into a nature-oriented Japan. Presented at the International Congress of Scientists on the Human Environment(HESC) at Kyoto in November 1975. 88 pages.

<P102> Shimoda,H.: Anti-IBM machines survived in '70s - Strategies of No.2 Burroughs in the 1980s, COMPUTOPIA, January 1980. Pp.54-63(in Japanese).

(P103) Shinn,J.: How will conflicts of LSI become in the future?, Nikkei-Business, March 1981. Pp.165-168(in Japanese).

(P104) Simon,H.A.: Rational decision making in business organizations, The American Economic Review, Vol.69, No.4, September 1979. Pp.493-513.

<P105> Simons,W.W.: Consenser, Technology and Economy, September 1980. Pp.85-94(translation in Japanese).

<P106> Special issues in new frontiers of computer science: Nishino, H.: New frontiers of computer science, COMPUTOPIA, April 1981. Pp.14-17(in Japanese); Takahashi,S,: How will computer systems change?, COMPUTOPIA, April 1981. Pp.18-24(in Japanese); Ishii,O.: New computer system and VLSI technology, COMPUTOPIA, April 1981. Pp.26-31(in Japanese).; Tanaka, H.: Understanding natural languages, COMPUTOPIA, April 1981. Pp.34-39(in Japanese).; Shirai,Y.: Artificial intelligence and robots, COMPUTOPIA, April 1981. Pp.50-53(in Japanese).

[P107] Sugiyama,K.: Dynamic simulation of environmental and economic behavior of the Japanese Islands: Terrestrial environmentology IV, J. Earth Sci. Nagoya Univ., Vol.21, No.1, 1973. Pp.1-42.

[P108] Sugiyama,K. and Shimazu,Y.: Some problems in economy-oriented environmentology: Terrestrial environmentology I, J. Earth Sci. Nagoya Univ., Vol.20, No.1, 1972. Pp.1-29.

<P109> Sugiyama,K., Tagawa,S. and Toda,M.: Methods for visual understanding of hierarchical system structures, IEEE Trans. SMC, Vol.SMC-11, No.2, February 1981. Pp.109-125.

<P110> Sugiyama,K., Tagawa,S. and Toda,M.: Visual representations of structural information - Algorithms for automatic drawing of hierarchical structures and their applications, Japanese Report No.2, International Institute for Advanced Study of Social Information Science, Fujitsu Ltd., 1981. 41 pages(in Japanese).

[P111] Suzuki,K.: A new development in R & D intensive middle scale enterprise, Technology and Economy, May 1981. Pp.4-18(in Japanese).

[P112] Symposium: Future issues in technological innovation in an age of stagnation, Technology and Economy, July 1979. Pp. 72-88(in Japanese).

(P113) Takagi,T.: Japanese science and technology policy from the viewpoint of diplomacy, Technology and Economy, June 1979. Pp.62-77(in Japanese).

[P114] Talk: Expectations for the new digital data communication network, COMPUTOPIA, January 1980. Pp.74-92(in Japanese).

<P115> Tanaka,H.: A topic on future computer systems, Denshi-kogyo-geppo, Vol.21, No.4, 1979. Pp.4-10(in Japanese).

(P116) Tanaka,R. et al.: A talk: Looking at the future of international trade problems among Japan, the U.S. and Europe, Bungei-shunju, August 1981. Pp.388-395(in Japanese).

(P117) Terasawa,Y.: IBM aiming at office automation, COMPUTOPIA, April 1981. Pp.54-66(in Japanese).

[P118] The editors of Technology and Economy: Information relating to basic researchers, Technology and Economy, October 1980. Pp.106-110(in Japanese).

[P119] The Special Committee for the Impacts of Microelectronics on Environment; A fact finding study on the impacts of microcomputers on employment - Summary, November 1979. 37 pages.

[P120] Tobioka,K.: Achieving a breakthough during stagnant situations, Technology and Economy, October 1980. Pp.18-39(in Japanese).

[P121] Tsuchiya,H.: Research on basic research, Technology and Economy, October 1980, Pp.4-17(in Japanese).

(P122) Uchida,H.: Excellent historical considerations relating to technology transfer, Kikan-gendai-keizai, Summer 1981. Pp. 186-189(in Japanese).

[P123] Uenohara,M.: Promotion of innovations in Japan. Presented at the 1980 International Conference on Innovation Policy held in Tokyo, Technology and Economy, June 1980. Pp.136-141(in Japanese).

(P124) Utterback,J.M.: Innovation in industry: The diffusion of technology, Science, Vol.183, No.4125. Pp.620-626.

(P125) Vasko,T.: Telecommunications: The Issues, IIASA Working Paper WP-82-1, January 1981. 10 pages.

(P126) Ward,E.P.: Focusing innovative effort through a convergent dialogue, Long Range Planning, Vol.13, No.6, December 1980. Pp.32-41.

(P127) Ward,E.P.: Planning for technological innovation - Developing the necessary nerve, Long Range Planning, Vol.14, No.2, April 1981. Pp.59-71.

<P128> Yamada,M.: The growth of personal computers in the 1980s, COMPUTOPIA, January 1980. Pp.94-99(in Japanese).

(P129) Yamadori,Y.: "A key to productivity" NCC'81, COMPUTOPIA, July 1981. Pp.48-54(in Japanese).

<P130> Yamaguchi,K.: Computer security and social security - Possibilities and limitations of computer systems as man-machine systems, COMPUTOPIA, March 1981. Pp.20-24(in Japanese).

[P131] Yamaguchi,U.: Problems in material resources for electronics, Denshi-kogyo-geppo, Vol.22, No.6, 1980. Pp.47-51(in Japanese).

(P132) Yanai,H.: I attended the workshop "microelectronics--new dimensional technological innovation and automation", Technology and Economy, January 1980. Pp.80-94(in Japanese).

<P133> Yasuda,J.: How far will the personal computer be developed in the 1980s?, COMPUTOPIA, March 1979. Pp.56-61(in Japanese).

(P134) Yershov,Y.V. and Saltysov,I.V.: On determinig future impacts of research and development solutions, Tech. Forec. Soc. Change, Vol.12, 1978. Pp.165-174.

(P135) Yokoyama,K.: On discussions about impacts of microelectronics on society in the U.S. and Europe, Technology and Economy, December 1979. Pp.65-83(in Japanese).

R & D Managment Systems in Japanese Industry
H. Eto and K. Matsui (editors)
© Elsevier Science Publishers B.V. (North-Holland), 1984

Chapter VI RECENT TRENDS IN R & D AND PATENTS --- A QUANTITATIVE APPRAISAL ---

KIMIO UNO

Institute of Socio-Economic Planning
The University of Tsukuba
Sakura, Ibaraki 305, Japan

This chapter discusses trends in R&D, followed by a discussion on the application and registrations of patents. Japan is spending about 2 percent of her GNP for R&D, and in absolute value Japanese R&D expenditures come after that of the United States, the Soviet Union, and perhaps Germany. The number of patents applied for is the largest in Japan. Patents can be considered to be a proxy of the output of R&D activities, and an attempt is made to relate the two in a series of regression analysis. It is revealed that Japan's elasticity of R&D expenditures to real GNP is 1.20 for 1965 - 1980 period, and that the annual growth of R&D expenditures has been 6.6 percent in real terms. It is also shown that the "efficiency" of production of technological knowledge is improving in Japan, in contrast to other Western countries where it is deteriorating. Japan seems to be successfully shifting from technological borrower to technological exporter, if we can assume patents to be a good indicator of invetiveness.

1. JAPAN SHIFTING FROM A TECHNOLOGY IMPORTER TO AN INNOVATOR

The R&D expenditures tend to increase for at least two reasons. First, with virtual elimination of the technology gap which existed between Japan and Western Europe and the United States, Japan can no longer rely on technology borrowing and is being required to spend more of her own resources in acquiring new technological knowledge. The technological shift toward electronics including

computer, communication, and control systems may be included in this category. Second, energy shortage, environmental pollution, limited availability of land, and other constraints require technological breakthroughs.
The portion of resources devoted to R&D has been on the increase both in terms of manpower and material inputs. The ratio of R&D expenditures to GNP is 1.96 percent in 1980, as compared to 2.33 percent for the United States, 2.19 percent for the U.K., 2.28 percent for Germany, and 1.80 percent for France.(1) The corresponding ratio is calculated to be 3.47 percent in the Soviet Union using the estimated GNP as the denominator.

R&D is not an end in itself. It represents the amount of effort being devoted to technological advancement, not the technological advancement itself. Although there is no universally accepted measure of the output of R&D, we may consider patents as a good proxy.

Admittedly, the purpose of R&D is not limited to acquisition of patents. R&D is commonly classified into basic, applied, and development research, and basic knowledge is perhaps not suitable for patents. It is said also that important technological knowhow is not made available even in the form of patents. Even with those reservations, if one can take the number of patents applied or granted as a proxy, this has been on the increase. As a matter of fact, Japan has become a leading country in this aspect, surpassing other industrialized countries. The number of patents applied for stood at 191,020 in 1980 (of which 165,730 were by the Japanese), or 382,805 including utility models.

The figure compares favorably with 104,329 for the United States, 48,583 (or 82,830 including utility models) for Germany, 41,640 for the U.K., 27,989 for France,and 168,589 for the Soviet Union. The number of patents granted is 46,106 in 1980 (or 96,107 including utility models), which is third after 94,582 for the Soviet Union and 61,827 for the United States.(2) The Japanese figure in the mid 1970s had been comparable to the Soviet one of around 45,000 cases, but the latter surged after 1978 whereas the former continued to hover at the same level. The U.S. figure has declined from about 70,000 in the mid 1970s.

Despite those success indicators, however, it is often pointed out that Japan is still heavily dependent on foreign technology and that the Japanese technological level is still relatively low compared to the technologically advanced West.

Technological balance of payments, which compares the technology related receipts and payments of a country, illustrates the point. Recent issue of the White Paper on Science and Technology, citing technological balance of payment statistics of the major industrial countries, reveals that Japan and Germany remain as technological net importers.(3) The United States is by far the largest technological exporter, selling 1,564.4 billion yen worth of technology abroad in 1981, followed by France (476.8 billion yen in 1977), U.K. (189.7 billion yen in 1979), Japan (118.4 billion yen in 1981), and Germany (107.7 billion yen in 1979). The biggest technology importer is Japan (377.2 billion yen in 1981); France stands second (332.8 billion yen in 1977), followed by Germany (233.4 billion yen in 1979), U.S. (194.7 billion yen in 1981), and the U.K. (147.3 billion yen in 1979). As far as technological balance of payments are concerned, the United States enjoys a huge surplus; as a matter of fact, her technological exports are nearly 9 times as large as her technological imports, as is shown in Table 1.

Table 1. Technological Balance of Payments in Selected Countries (Ratio, receipts/payments)

Year	Japan			U.S.	U.K.	France	Germany
	(A)	(B)	(B)				
1965	0.10	n.a.	(n.a.)	11.36	1.06	0.79	0.45
1970	0.14	'71 0.20	('71 0.71)	10.36	1.07	0.96	0.39
1975	0.23	0.39	(1.42)	9.09	1.02	1.27	0.42
1980	0.26	0.67	(2.68)	8.93	'78 1.29	'77 1.43	'78 0.46

Note : The table shows the ratio of receipts to payments. For Japan, (A) is from Bank of Japan statistics, (B) from Bureau of Statistics, Prime Minister's Office statistics. Japanese figures in parentheses refer to new contracts only.

Source : Science and Technology Agency, White Paper on Science and Technology; ---, Indicators of Science and Technology.

Japanese receipts is less than 30% of her payments according to the Bank of Japan statistics, or about 67% according to statistics compiled by the Bureau of Statistics.(4) However, one should note that, when it comes to newly contracted technological trade, Japanese exports amount to 2.7 times her imports. We should interpret those figures as showing that existing technological contracts mainly reflect the large gap which existed until recently and that technological balance of trade largely reflects the stock of technological knowledge rather than new addition to it.

A recently conducted survey by the MITI Bureau of Industrial Technology (MITI, 1982) compares Japanese technological level with that of the U.S. and one more country considered most advanced in respectively fields. Some 160 key technologies were identified which are then grouped into the following 10 categories.

(1) Material technology:
 - (a) Technology for new material development
 - (b) Technology for material processing

(2) Fabrication-assembly technology:
 - (c) Technology for scaling up
 - (d) Technology for automation
 - (e) Technology for high-function processes
 - (f) Technology for testing and inspection
 - (g) Technology for production management

(3) Product technology:
 - (h) Technology for high-function product development
 - (i) Software technology
 - (j) Design technology

The result is summarized in Table 2. M. Usui, citing the survey, observes that the Japanese technological advantage is most readily visible in such domains as 'production management' and 'material processing' and that her disadvantages are being felt in the development of new materials and products, especially those meeting highly specialized requirements, as well as in design technology in general (Usui 1983). Thus he confirms the generally accepted view that Japan has improved her international competitiveness by achieving a high level of efficiency and quality control for the products originally developed elsewhere, and with timely 'advance' investment to reap the merits of standardization and the economies of scale.

Table 2. Comparative Technological Position of Japan by 10 Categories of Key Technology (as of 1982)

	Material Technology		Fabrication-assembly Technology					Product Technology			Total
	(a)	(b)	(c)	(d)	(e)	(f)	(g)	(h)	(i)	(j)	
Japan is:											
Superior to:											
U.S.	6	4	2	6	12	1	6	13	2	2	54
Europe	7	2	4	7	18	-	5	13	1	6	63
Equal to:											
U.S.	2	1	1	6	18	2	3	23	2	2	60
Europe	5	3	2	6	16	3	3	28	2	4	72
Inferior to:											
U.S.	8	-	4	4	7	2	1	22	4	20	72
Europe	2	-	1	2	2	2	-	10	-	10	30
Total, relative to:											
U.S.	16	5	7	16	37	5	10	58	8	24	186
Europe	14	5	7	15	36	5	8	52	3	20	165

Source: MITI, Bureau of Industrial Technology, 1982. Cited from Usui, 1983.

These two observations, one based on the technological balance of payments and the other on the comparative technological position of Japan, are an accurate description of today's technological level in Japan. We should note, however, that by referring to the level, we are referring to the stock of technological knowledge, rather than the new acquisition of the knowledge. In this chapter we are concerned with the latter. We can justify this approach by saying that the current efforts and achievements are better measured by current flow, and that future stock is determined by the magnitude of the current flow.

The observation that Japan still lags in some technology areas is not contradictory to the fact that she is the leading country in the production of technological knowledge: the former refers to the accumulated stock of knowledge whereas the latter refers to the new addition to it. In fact, according to the technological balance of payments statistics which reflect the stock of technological knowledge, Japanese technological trade has turned into surplus as far as new contracts are concerned. Evidently, new contracts more or less reflect the new technological knowledge being obtained.

In the following, trends in R&D expenditures in Japan are explained with empirical data. Then, trends in patent application and registration are discussed. Finally, R&D and patent trends are related in a series of regression analysis. International comparisons are attempted when considered appropriate.

2. TRENDS IN RESEARCH AND DEVELOPMENT

R&D is a broad concept. The Bureau of Statistics carries out the Survey of Research and Development annually, which defines R&D as "systematic studies and creative efforts in science and technology which are undertaken for the advancement of knowledge about natural phenomena, or for the development and technical improvement of products and production systems, etc." It includes research undertaken in research institutions as well as pilot studies and experiments in pilot plants in factories. In the following we will use this definition.

To be more precise, R&D activities include planning, designing, data collection, experiments, testing, inspection, analysis, reporting and so on. Included also are construction of necessary machinery, utensils, equipment, etc., rearing of animals and plants, and study of reference documents. In the case of production units, designing, construction, and testing of pilot plants and proto-type models are also included. Accounting and other general services for R&D are also included.

The following activities are not considered here as R&D. Quality control and routine examination, testing, and analysis of products, soil, or atmosphere; designing of machinery, utensils, or equipments for products beyond R&D activities using pilot plants or prototype models; geological and geophysical survey and mapping, exploration of underground resources, and the collection of general data on geological survey, ocean exploration, or astronomical observation and the like.

We should focus our attention on research in natural science fields. Table 3 shows the trends in R&D expenditures in Japan from fiscal year 1965 to 1980.[5] Column 2 indicates that in 1980, the R&D in natural science amounted to 4683.8 billion yen, which, according to column 3, is slightly less than 2 percent of the gross national product. According to column 4 which indicates the time series figures in 1975 prices, R&D expenditures have increased from 1129.4 billion yen in 1965 to 3389.1 billion yen in 1980 after adjusting for the changes in price levels, or an increase of three-fold in a fifteen year period. Average annual rate of increase stood at 15.2 percent for the 1965-1970 period, 2.9 percent for the 1970-1975 period, and 5.3 percent for 1975-1980 period.(6)

Table 3. R&D Expenditures in Japan

Fiscal Year	Gross National Product (billion yen)	R&D Expenditures (billion yen)	Share in GNP (percent)	R&D Expenditures, 1975 Prices (billion yen)	Annual Growth of Real R&D (percent)
1965	33550.2	425.8	1.27	1129.4	
1966	39452.0	488.7	1.24	1227.9	
1967	46175.6	606.3	1.31	1447.0	15.2
1968	54689.2	767.8	1.40	1729.3	
1969	64850.8	933.2	1.44	1940.1	
1970	75091.6	1195.3	1.59	2281.1	
1971	82725.8	1345.9	1.63	2399.1	
1972	96424.0	1586.7	1.65	2580.0	
1973	116636.3	1980.9	1.70	2669.7	2.9
1974	138044.6	2421.4	1.75	2617.7	
1975	151797.0	2621.8	1.73	2621.8	
1976	170290.0	2941.4	1.73	2698.5	
1977	188804.3	3233.5	1.71	2809.3	5.3
1978	206792.5	3570.0	1.73	3005.1	
1979	221965.6	4063.6	1.83	3174.7	
1980	239154.8	4683.8	1.96	3389.1	

Note: R&D expenditures in this table refer to intramural expenditures on R&D by the institutions mainly engaged in research in natural science fields; those by humanity and social science fields are not covered.

Data: Bureau of Statistics, Report on the Survey of Research and Development; Economic Planning Agency, Annual Report on National Accounts. R&D deflator from Science and Technology Agency, White Paper on Science and Technology 1982, p.438.

Source: Science and Technology Agency, Indicators of Science and Technology, 1982, p.28.

The rate of increase of R&D expenditures has been faster than that for GNP, resulting in the rising share of R&D in total GNP. In 1965, Japan was spending 1.27 percent of her total product on R&D, which rose to 1.59 percent in 1970, 1.73 percent in 1975, and 1.96 percent in 1980.

How does this compare with other major countries? Table 4 gives some clue to this interesting question. The table shows R&D expenditures in the United States, Soviet Union, United Kingdom, France, and Federal Republic of Germany, as well as those in Japan. The figures are expressed in Japanese yen in order to facilitate comparison. However, international comparisons are always hazardous and a word of caution is in order. First, GNP and R&D expenditures in the selected countries have been converted to Japanese yen using exchange rates which were effective in the respective years. Exchange rates are determined largely in the foreign trade sector, with increasingly larger influence exerted by international capital transactions. They may not be true indicators of relative purchasing power of the currencies involved, since many economic activities are not traded in the international market. Second, the definition of R&D may differ from one country to another. R&D expenditures by the U.S., Soviets, and France include humanities and social science fields.

With those limitations in mind, we turn to R&D expenditures in Table 4. According to the estimates, the United States is spending the largest sum, followed by the Soviet Union. Germany seems to be leading Japan by a slight margin in 1980. In 1965 Japan was the last in the six countries listed. In 1965, Japanese R&D expenditures were less than 6 percent of the U.S.; in 1970 about 13 percent; in 1970 about 25 percent; and in 1980 about 34 percent. If we exclude defense related research from both countries, the ratio narrows to about 43 percent. The shares of R&D expenditures in GNP are included in the same table. For Japan, two kinds of shares have been calculated: one excluding humanities and social sciences, and the other including those. As we have noted above, the figures for the U.S., Soviet Union, and France do include those fields.

The Japanese figure of about 2 percent (or 2.2 percent including social sciences) is not particularly high among the countries listed. In fact, it comes after the Soviet Union, Germany, the United States, and perhaps the United Kingdom, for which the 1980 figure is not available. We notice from the table that the shares of resources devoted to R&D have declined in the United States, France, and perhaps United Kingdom during the fifteen year period. In contrast, the shares have increased in Germany, Japan, and the

Table 4. R&D Expenditures in Selected Countries

Year	Japan	U.S.	U.S.S.R.	U.K.	France	Germany
GNP in comparable currency unit				(billion Japanese yen)		
1965	33,550	248,748	97,252	36,429	36,123	41,467
1970	75,092	351,896	144,279	44,158	50,762	66,691
1975	151,797	459,803	186,691	69,685	100,737	124,840
1980	239,155	595,599	211,142	118,962	148,762	186,115
R&D Expenditures in comparable currency unit (billion Japanese yen)						
1965	426	7,246	2,772	'66 911	730	716
1970	1,195	9,339	4,655	929	982	1,454
1975	2,622	10,438	6,885	1,410	1,813	2,293
1980	4,684	13,864	7,319	'78 1,463	'79 2,241	4,899
R&D Expenditures, as percentage of GNP						(%)
1965	1.27(1.52)	2.91	2.85	'66 2.35	2.03	1.73
1970	1.59(1.81)	2.65	3.23	2.10	1.93	2.18
1975	1.73(1.96)	2.27	3.69	2.02	1.80	2.38
1980	1.96(2.19)	2.33	3.47	'78 2.19	'79 1.77	2.63
R&D Expenditures, current prices, in national currencies						
	(billion yen)	(billion dollars)	(billion rubles)	(million pounds)	(billion francs)	(billion marks)
1965	425.8	20.0	6.9	'66 0.9	9.9	7.9
1970	1,195.3	26.1	11.7	1.1	15.2	14.8
1975	2,621.8	35.2	17.4	2.1	26.2	24.6
1980	4,683.8	61.1	21.3	'78 3.6	'79 43.5	39.3
R&D Expenditures, 1975 prices, in national currencies					(unit same as above)	
1965	1,129.5	33.8	6.9	'66 2.0	19.1	12.9
1970	2,281.2	35.8	11.7	2.0	23.1	20.3
1975	2,621.8	35.2	17.4	2.1	26.2	24.6
1980	3,389.1	43.0	21.3	'78 2.5	'79 30.1	32.5

Note: Comparable price figures were obtained by converting by foreign exchange rates in respective years. Japanese figures in parenthesis include humanities and social sciences. Soviet GNP series is from NSF, Science Indicators 1980. U.S., Soviet, and French R&D figures include humanities and social sciences. 1975 price figures were estimated by deflating current price series by implicit GNP deflators for each country obtained from International Financial Statistics; for Japan, R&D expenditure deflator estimated by the Science and Technology Agency was employed. No adjustment was made for the Soviet data for lack of appropriate price data; official statistics reports that industrial whole price index (1949=100) has risen three percentage points between 1965 and 1978.

Source: Science and Technology Agency, White Paper on Science and Technology.

Soviet Union. However, annual figures indicate that the shares of R&D are increasing in all countries in recent years.

Table 5. R&D Personnel in Japan

(thousand persons, %)

	Total	Researchers		Assistant Research Workers		Technicians		Clerical and others	
1965	303.8	117.6	(38.7)	75.9	(25.0)	62.6	(20.6)	47.6	(15.7)
1966	323.0	128.9	(39.9)	82.9	(25.7)	60.2	(18.6)	51.0	(15.8)
1967	327.6	138.7	(42.3)	80.1	(24.4)	61.2	(18.7)	47.7	(14.6)
1968	356.3	157.6	(44.2)	83.2	(23.3)	68.2	(19.1)	47.3	(13.3)
1969	367.3	157.1	(42.8)	70.8	(19.3)	84.6	(23.0)	54.9	(14.9)
1970	392.2	172.0	(43.9)	75.4	(19.2)	88.3	(22.5)	56.6	(14.4)
1971	429.3	194.3	(45.3)	80.2	(18.7)	93.6	(21.8)	61.2	(14.3)
1972	426.9	198.1	(46.4)	82.3	(19.3)	86.1	(20.2)	60.4	(14.1)
1973	459.2	226.6	(49.3)	80.7	(17.6)	88.9	(19.3)	63.1	(13.7)
1974	468.1	238.2	(50.9)	79.4	(17.0)	87.5	(18.7)	63.0	(13.5)
1975	491.3	255.2	(51.9)	81.9	(16.7)	90.6	(18.5)	63.5	(12.9)
1976	488.0	260.3	(53.3)	79.2	(16.2)	89.1	(18.3)	59.4	(12.2)
1977	492.3	272.0	(55.2)	73.8	(15.0)	86.7	(17.6)	59.8	(12.2)
1978	486.8	273.1	(56.1)	72.5	(14.9)	83.3	(17.1)	57.9	(11.9)
1979	496.0	281.9	(56.8)	73.0	(14.7)	82.2	(16.6)	59.0	(11.9)
1980	521.1	302.6	(58.0)	73.9	(14.2)	85.9	(16.5)	58.7	(11.3)

Note: As of April 1 of the year stated.
Data: Bureau of Statistics, Report on the Survey of Research and Development.

Another major resource input for research is of course human resources. Table 5 shows R&D personnel in Japan by occupational category, which include researchers, assistant research workers, technicians, and clerical and other occupations. Researchers are defined as those who have a university degree or an equivalent and who have research experience of two years or more, and are engaged in research activities with their own themes. Included in the table are regular researchers, or persons who are engaged mainly in intramural research activities. Assistant research workers are those who assist researchers and who are engaged in research activities under their directions. Technicians are those engaged in testing and inspection for researches, or those engaged in manufacturing of machinery, tools, and equipment for research. Clerical and others include those who are engaged in miscellaneous business, general affairs, accounting, etc., and laborers such as guards, drivers, etc., related to research activities.

According to the table, there are 521 thousand persons in 1980 who are engaged in R&D related occupations. Of these, 58 percent are researchers. The percentage share of researchers has increased from about 40 per cent in the mid sixties. The annual rate of increase is calculated to be 7.9 percent for 1965-1970, 8.2 percent for 1970-1975, and 3.5 percent for 1975-1980.

Table 6. Researchers in Selected Countries

(thousand persons)

Year	Japan	U.S.	U.S.S.R.	U.K.	France	Germany
1965	117.6	494.5	664.6	'67 56.6	44.0	61.6
1970	172.0	540.9	927.7	54.7	57.3	78.4
1975	255.2	533.1	1223.4	78.8	62.0	94.1
1980	302.6	643.5	1373.3	'78 104.4	'79 72.9	122.0

Note: Figures for U.K., France, and Germany are partly estimated by interpolation.
Source: Science and Technology Agency, White Paper on Science and Technology.

Table 6 lists the number of researchers in selected countries. The Japanese figures refer to researchers only. The Japanese figure for 1980 stands at 303 thousand persons, which is about 47 percent of the U.S. figure of 644 thousand persons. Unlike R&D expenditures, Germany seems to have fewer researchers than Japan. The Soviet Union seems to have by far the largest number of researchers, totaling more than 1373 thousand. Here again, we are not too sure about the consistency of definition among countries. Nevertheless, trends within each country give us meaningful information since they do not suffer from a definitional problem. During the 1965-1980 period, the largest increase in the numer of researchers is recorded in Japan. The number of Japanese researchers increased 2.6 times in the fifteen year period, followed by 2.1 times in the Soviet Union, and 2.0 times in Germany. The United Kingdom and France seem to show a considerable increase, although the picture is somewhat confused by the limited availability of statistics. In the United States, the number of researchers increased from 495 thousand in 1960 to 644 thousand in 1980, an increase of 1.3 times.

3. TRENDS IN PATENT APPLICATION

The number of patents applied for in Japan registered 191,020 in 1980. Besides patents, the Japanese industrial ownership system has utility models, application for which numbered 191,785 in 1980[7]. Table 7 summarizes the number of patents applied for and granted in Japan since 1965. In 1965, patent application numbered 81,923 cases; for 1980 the figure is more than twice as large.

Table 7. Patents Applied for and Granted in Japan

(Number of cases)

Year	Patents Applied: Total	by Japanese (%)	Patents Granted: Total	by Japanese (%)
1965	81923	60796 (74.2)	26905	17793 (66.1)
1966	86046	62962 (73.2)	26315	17373 (66.0)
1967	85364	61721 (72.3)	20773	13877 (66.8)
1968	96710	71114 (73.5)	27972	18576 (66.4)
1969	105586	77132 (73.1)	27657	18787 (68.1)
1970	130831	100522 (76.8)	30878	21390 (69.3)
1971	105785	78425 (74.1)	36447	24795 (68.0)
1972	130400	101328 (77.7)	41454	29101 (70.2)
1973	144814	115221 (79.6)	42328	30937 (73.1)
1974	149319	121509 (81.4)	39626	30873 (77.9)
1975	159821	135118 (84.5)	46728	36992 (79.2)
1976	161016	135762 (84.3)	40317	32465 (80.5)
1977	161006	135991 (84.5)	52608	43047 (81.8)
1978	166092	141517 (85.2)	45504	37648 (82.7)
1979	174569	150623 (86.3)	44104	34863 (79.0)
1980	191020	165730 (86.8)	46106	38032 (82.5)

Source: Patent Agency, Patent Agency Annual Report.

Patent applications include those by foreign nationals; application by the Japanese was 74 percent in 1965, which rose steadily throughout the period reaching 77 percent in 1970, 85 percent in 1975, and about 87 percent in 1980.

The increasing trend in the patent application was not smooth. One notices a dip in the number of applications in 1971, which is attributed to a revision in the Patent Law in that year. Rather stagnant patent application during the 1975-1977 period is a reflection of sluggish economic performance following the Oil Crisis in the fall of 1973. Patent application seems to have revived only after 1978.

Not all applications actually materialise as patents. The number of patents granted in 1965 totaled 26,905, of which 66 percent were to Japanese nationals. In 1980, total number of patents granted rose to 46,106, of which 83 percent were to the Japanese. Due to the time lag between application and registration, the application dip in 1971 was felt in 1974, or three years later. Likewise, the slowdown until 1977 is still being reflected in the number of patents granted. We can predict, based on our previous observation on the time lag, that the trend will turn upward in the 1980s.

Comparing the shares of patent application and registration by the Japanese, one notices that the former is consistently larger than the latter. One should not hastily conclude that the Japanese applications tend to be turned down at a higher rate. Judging from the time lag, it would be more appropriate to compare the two series with about a three year lag. This considerably narrows the difference between the number of patents applied for and granted, although there remains some difference.

Among major industrial countries, patent application in Japan has been the largest in number since 1970. In 1969, the Soviet figure was the largest. The Soviet patent system is different from those in the market economies in that it mainly grants an inventor's certificate and patents are rather exceptional.[8] This peculiar system is due to the fact that all enterprises are nationalized; being such, invention belongs to the nation. Excluding the Soviet Union, application in the United States had been the largest until 1967.

Table 8 summarizes patent application in selected countries. In some countries, i.e., U.K., France, and Germany, the number of applications has been declining, especially in the 1970s. Japan, the United States, and the Soviet Union have recorded increases during the period under consideration. However, if we exclude application by foreign nationals, the United States also shows a declining trend since 1970.

Table 8. Patents Applied for in Selected Countries

Year	Japan	U.S.	U.S.S.R.	U.K.	France	Germany
Patent Application Total					(number of cases)	
1965	81923	94629	103616	55507	47793	66470
1970	130831	103175	115520	62101	47283	66132
1975	159821	101014	119163	53400	40437	60095
1980	191020	104329	168589	41640	27989	48583
Patent Application by Own Nationals					(number of cases)	
1965	60796	72317	102624	24724	17509	38148
1970	100513	76195	110574	25227	14166	32772
1975	135118	64445	114488	20842	12110	30198
1980	165730	62098	165459	19612	11000	28683
Patent Application by Foreign Nationals					(number of cases)	
1965	21127	22312	992	31233	30284	28322
1970	30318	26980	4946	36874	33117	33360
1975	24703	36569	4708	32558	28327	29897
1980	25290	42231	3130	22028	16989	19900
Share of Application by Foreign Nationals						(percent)
1965	25.8	23.6	1.0	56.3	63.4	42.6
1970	23.2	26.2	4.3	59.4	70.0	50.4
1975	15.5	36.2	4.0	61.0	70.1	50.0
1980	13.2	40.5	1.9	52.9	60.7	41.0

Note: Soviet figures include "inventor's certificate".
Source: Patent Agency, Patent Agency Annual Report.

Patent application by a country's own nationals can be considered to be an indicator of endogenous innovative effort. In 1980, Japan ranks first with 165,730 application by the Japanese, followed by the U.S. with 62,098, Germany 28,683, U.K. 19,612, and France 11,000.

The share of patent application by foreign nationals is the lowest in Japan where 13 percent is attributed to foreigners. The ratio has halved from about 26 perent in 1965. In the United States, the share of application by foreigners has steadily increased from 24 percent in 1965 to about 40 percent in 1980. The situation in European countries is mixed; up until the mid 70s, foreign shares had been on the increase, but this has since been reversed.

With the exception of the Soviet Union where the economic system differs greatly from the market economies, the number of applications by foreign nationals does not differ very much among countries. This can be interpreted as a reflection of the fact that most major inventions are being registered in major markets simultaneously.

Table 9. Patents Granted in Selected Countries

(Number of cases)

Year	Japan	U.S.	U.S.S.R.	U.K.	France	Germany
1965	26905	62857	13158	33864	41800	16780
1970	30878	64427	32466	40995	26297	12887
1975	46728	71994	44121	40689	14320	18290
1980	46106	61827	94582	23804	28060	20188

Note: See Table 5.
Source: See Table 5.

The trends of the number of patents granted gives a slightly different picture than the one we have seen above. According to Table 9 which lists time series data for selected countries, the United States leads other market economies with 61,827 patents granted in 1980. Japan follows the United States with 46,106 cases. Considering the shares of application by foreign nationals in the two countries, however, domestic R&D efforts seem to be yielding nearly the same result as far as patents are concerned. Among European countries, Germany appears to show an upward trend since 1970: if one examines annual data, however, this is not so. In 1969, for example, Germany registered 22,623 patents, and 23,934 cases were registered in 1973. The Soviet Union exhibits a much steeper trend in the number of registrations than applications.

In any case, one should be cautions about the international comparisons of patent numbers, because they reflect the differences in the industrial ownership system. It would be more meaningful to attempt time series analysis on the efficiency of the technological knowledge production within each country, to which we turn in the next section.

4. PRODUCTION OF TECHNOLOGICAL KNOWLEDGE

We have discussed the trends in research and development expenditures in section 2 and the trends in patent application in section 3. In this section, we shall try to relate the two in order to gain some insight into the efficiency of R&D activities. In other words, we shall try to estimate the production function of technological knowledge. We proceed by assuming that the number of patent applications is a proxy of the output of R&D.

Productivity of knowledge production is important in judging the future trend of growth of the Japanese economy. In the short run, the macro economic effect of R&D expenditures will be felt mainly on the demand side, i.e., as a component of the effective demand. In the long run with sufficient time allowed for the new technological knowledge to be embodied in new modes of production, the effect of R&D will be felt on the supply side. In the case of improvement in the products, the effect of R&D will be felt through quality improvement or cost reduction, thereby influencing the demand structure of the economy.

Let us begin by summarizing the trends in R&D expenditures. Following is the result of regression analysis which explains R&D expenditure by gross national product. Both terms are in logarithm, so that the obtained parameter shows the elasticity of R&D expenditures to GNP. The elasticity in this case may be interpreted as an indicator of the R&D efforts in relation to the available resources as measured by GNP. The functional form is:

$$\log (Y) = a + b \log (X) \quad \ldots\ldots\ldots\ldots\ldots\ldots \quad (1)$$

where X : Gross national expenditure

Y : R&D expenditures

The calculated parameters for the countries under consideration are as follows.

Japan	(1965-1980) :	1.2020 (t value = 70.2)
United States	(1965-1980) :	0.7739 (t value = 27.2)
Soviet Union	(1965-1980) :	1.2675 (t value = 27.8)
United Kingdom	(1966-1978) :	0.9535 (t value = 38.2)
France	(1965-1979) :	0.8761 (t value = 55.0)
Germany	(1965-1980) :	1.2657 (t value = 37.6)

The elasticity is estimated to be about 1.20 for Japan for the period 1965 to 1980, implying that as the economy grows 1 percent, R&D expenditures would grow 1.2 percent. Apparently, Japan has been pouring resources into R&D activities more than proportionately than for other purposes.

The Japanese figure, however, is not the highest among the countries considered here. Both Germany and the Soviet Union recorded about 1.27. The elasticities are estimated to be less than unity in the United States, the United Kingdom, and France, of which the United States shows the lowest figure of 0.77.

The growth rates of GNP are different among countries; therefore, the elasticity of R&D expenditure alone does not tell us very much about the actual growth rates of R&D expenditure. By fitting a time trend variable to the logarithmic form of R&D expenditures in real terms (1975 prices), we obtain the average annual growth rates of R&D in respective countries. The functional form is:

$$\log (Y) = a + b\,T \quad \text{.....................} \quad (2)$$

where T : time trend variable
Y : R&D expenditures

The result is summarized below.

Japan	(1965-1980) :	0.0655	(t value = 9.6)
United States	(1965-1980) :	0.0099	(t value = 4.0)
Soviet Union	(1965-1980) :	0.0770	(t value = 17.0)
United Kingdom	(1966-1978) :	0.0163	(t value = 4.9)
France	(1965-1979) :	0.0276	(t value = 18.3)
Germany	(1965-1980) :	0.0567	(t value = 14.4)

Ignoring for now the kinks in the trends during the period under consideration, average rate of increase of real R&D expenditures in Japan is calculated to be approximately 6.6 per cent a year. This is exceeded by the Soviet Union, where the rate of increase was 7.7 percent a year. The rate of increase was relatively high in Germany, where it is calculated to be 5.7 percent.

Care must be taken, however, in interpreting the figures presented above because the rate of increase may differ from period to period reflecting various internal and external factors. In Japan, for example, the annual rate of increase of R&D expenditure after adjusting for inflation varied from 15.2 percent for 1965-1970 to 2.9 percent for 1970-1975, which then recovered to 5.3 percent for 1975-1980 as we have seen in Table 3 above. This reflects the rather high rate of economic growth in the sixties, turbulent years following the Oil Crisis in 1973, and stable growth in recent years. The figures summarized above ignore such shifts in the trends and give the overall picture during the fifteen year period from 1965 to 1980.

An interesting question is how R&D expenditures relate to patent applications. The number of patents applied for in a country refers to those by a country's own nationals in this paper because our purpose is to see the R&D efforts and the outcomes. Both the number of applications for patents and R&D expenditures are in logarithm so that the result shows the elasticity. The function is of the following form.

$$\log (Y) = a + b_1 \log (X) + b_2 T \qquad (3)$$

where T : time trend variable

X : the amount of R&D expenditures per research worker

Y : the number of patents applied for per research worker

Dummy variables were employed case by case in order to remove the influence of exogenous factors, e.g., the revisions of the Japanese Patent Law in 1971 which resulted in a rather sharp drop in patent applications that year.

The result of the regression is reported below. Form A refers to the case where the functional form (3) above was found statistically significant. Form B is an alternative where only the trend variable was employed in explaining the number of patents applied for per research worker. Although the former is the preferred form, it resulted in a wrong sign or insignificant parameters for some countries, in which case only the form B is reported.(9)

Japan (1965-1980)	: (A)	0.25941 log X	(t value = 1.56)
		0.00594 T	(t value = 1.78)
	(B)	0.00667 T	(t value = 1.92)
United States (1965-1980)	: (B)	-0.02103 T	(t value = 5.46)

Soviet Union (1965-1980)	: (A)	0.26118 log X	(t value = 8.77)
		0.04785 T	(since 1974; see note)
	: (B)	-0.02658 T	(t value = -3.78)
United Kingdom (1967-1978)	: (A)	0.86209 log X	(t value = 4.09)
		-0.05274 T	(t value =-4.18)
	(B)	-0.09744 T	(t value = 9.67)
France (1965-1979)	: (B)	-0.06455 T	(t value =-13.1)
Germany (1967-1979)	: (B)	-0.06225 T	(t value =-15.7)

The result for Japan seems plausible. Research effort as represented by the amount of R&D expenditures per researcher, together with the improvement in the efficiency of "knowledge production", explains the number of patents applied for per researcher. The efficiency improvement is estimated to be at the rate of about 0.59 percent per year according to form A and 0.67 percent according to form B. The elasticity of technological knowledge production to resource input is estimated to be 0.259. In the case of form A, the fitted function explains about 63 percent of the variation.

The functional form A was found empirically meaningful for the Soviet Union and the United Kingdom.(10) The estimated elasticity of 0.261 for the Soviet Union is rather similar to that of Japan which we have seen above. The United Kingdom seems to have much higher elasticity of 0.862; the efficiency of technological knowledge production, however, is decreasing at a high rate, as we see in a rather large negative coefficient on the time trend variable.

Actually, all countries under consideration are experiencing deterioration in efficiency except Japan. Our results indicate the annual gains in efficiency to be -2.1 percent for the United States, -2.7 percent for the Soviet Union, -9.7 percent for the United Kingdom, -6.5 percent for France, and -6.2 percent for Germany. In the case of the Soviet Union, however, we have observed gains in efficiency at an annual rate of 4.8 percent in recent years.

In summary, our attempt to explain the number of patent applications by the amount of R&D expenditures, both per researcher, was only partly successful. It gave a reasonably good result for

Japan, suggesting that in this country R&D effort is directly related to the "production of technological knowledge" as measured by the number of patents applied for. The same is true in the Soviet Union and the United Kingdom. In other countries, however, resource inputs in R&D do not seem to be reflected in the number of patents. In the United States, France, and Germany, only the trend variable was found significant; a negative sign on the trend variable indicates that the production of patents is becoming less efficient. In the United Kingdom, although the resource inputs do explain the patent application, a declining trend in efficiency seems to dominate the overall performance.

5. FURTHER REMARKS

In this chapter, we first examined the trends in research and development expenditures in Japan, and found that the rate of increase of R&D expenditures has been faster than that of GNP and, as a result, the share of R&D in the national economy has been rising to reach about 2 percent (2.2 percent when we include humanities and social sciences).

Second, we discussed the trends in patent application. The number of patents applied for in Japan has been the largest in number among major industrial countries. The number of patents actually granted lags behind application but the increasing trend is evident.

Third, an attempt was made in the previous section to explain the number of patent applications by the R&D expenditures. In other words, assuming that the patent application would be a proxy of the volume of scientific knowledge being obtained by R&D effort, production of technological knowledge was quantitatively analyzed.

A series of regression analysis was attempted which gave a mixed result as to the productivity of R&D if we are to assume that the number of patents applied for is a good proxy of the output of R&D activities. However, the obtained result seems to indicate that the production of technological knowledge in Japan has been rather efficient in the sense that the amount of R&D expenditures is directly reflected in the number of patents applied for. For other countries, research effort and the number of patents do not seeem to be directly related.

This leaves us with at least two alternative interpretations. One is that the efficiency of technological knowledge production is good in Japan and bad elsewhere. Second is that the research effort in major industrial countries, with the exception of Japan, is directed toward areas where the outcome cannot be appropriately measured by the number of patents being applied for. If the latter is the case, the number of patents is not a good indicator of technological knowledge being obtained through R&D activities.

In fact, the main focus of the research and development in the United States is in such areas as space technology, ocean development, and nuclear energy development. We should perhaps add defense to the list. Compared with the U.S. level, the Japanese expenditure in space technology is only 10 percent, that for ocean development 25 percent, and that for nuclear energy 37 percent; judging from the fact that the overall R&D expenditure in Japan is about 34 percent of the U.S. level, the difference in the resource allocation pattern is evident. Needless to say, they are the areas where major technological breakthroughs are being made. One of the characteristics of those areas is that they are system oriented. Being such, research and development in those areas may not be aimed at acquiring knowledge suitable for patent registration.

One can argue further that such technological knowledge, although valuable in itself, may not be embodied in products which can be easily exported to the world market. In contrast, Japanese R&D effort has focused on industrial knowledge which can be transformed into products which can be traded. Thus, the Japanese were able to shift from transistor radios, motor cycles, and steel to video tape recorders, industrial machinery, and automobiles. Next in line are various electronic products including computers, industrial robots, and communication equipment.

In the export market as well as in the domestic market, technological innovation embodied in the products often make them unique. In the aggregate the goods in question may fall under a particular commodity classification which are produced rather universally; but in reality they are new products. Or one can say that they become differentiated in the eyes of the purchasers. This means that in explaining the trade, availability theory rather than comparative advantage theory becomes more relevant.(11) This implies further that relative prices including the effect of floating exchange rates become less influential in adjusting the trade flow.

It is said that the Japanese technology, given the fact that Japan has been a follower in this field, is weak in basic research and that the Japanese contribution is limited to minor improvements to the existing technology which originated elsewhere. It is difficult to assess the overall quality of a country's new technology, apart from case by case comparisons of a particular technology or product. Suffice it to say that, judging from the capital labor ratios in Japanese industries, the new technology being developed in this country is not of a minor importance. When one compares the capital labor ratios in Japan and the United States, one notices that the differences were narrowed and eventually reversed in the late 1960s or in early 1970s in most manufacturing industries. Today, Japanese steel workers are equipped with nearly four times as much capital stock as their American counterpart; auto workers three times as much.(12) Although in the economy as a whole, capital labor ratio is higher and productivity higher in the United States, we may say that many of the Japanese industries are on the technological frontier, forcing them to be innovative. No single country has been the sole source of technological knowledge. It is true, however, that there are better chances of new technology actually being implemented where the rate of investment is relatively high. It is a historical fact that a fast growing nation absorbs new knowledge being obtained in technologically advanced countries, and Japan was no exception.

For the past three decades, the main source of technological knowledge has been the United States. Reflecting the huge technological gap which existed between the two countries, the application of patents by the Americans in Japan far exceeded the application by the Japanese in the United States. In recent years, however, the flow of technological knowledge between the two countries is more or less in balance. In 1960, the flow of technological knowledge in the form of patent application from Japan to the United States was less than 15 percent of the American application in Japan. Even in 1970, the Japanese application in the United States stood at about 50 percent of the flow in the opposite direction. In 1975, the flow from Japan exceeded the flow from the United States for the first time. This is mainly due to the increase in the flow from Japan; but one cannot ignore the fact that the application from the United States has declined from about 13,000 cases a year around 1970 to about 10,000 cases in recent years.

If the number of patents is a good indicator of the volume of technological knowledge being acquired, we may call the Japanese R&D policy a success. The question remains, however, if this is really so, when we consider the increasing importance of system orientation as we observe in newly expanding technological frontiers.

Note:

(1) The U.S. figure is an estimate for 1980. Figures for the U.K., Germany, and France are for 1978, 1979, and 1979, respectively. The Soviet figure is for 1979. The ratio appears high partly because NMP concept is narrower than GNP. Science and Technology Agency, ed., Indicators of Science and Technology, 1982 edition.

(2) Patent Agency, Patent Agency Annual Report

(3) Science and Technology Agency, White paper on Science and Technology, 1982 edition, pp.233-240.

(4) The two sources differ in the following: (a) The Bank of Japan statistics cover foreign exchange transactions for the purpose of technological assistance only; it excludes, for example, technology export accompanying plant exports. (b) The Bank of Japan statistics cover all transactions, whereas the Bureau of Statistics figures cover the business firms engaged in R&D activities; this results in the exclusion of nonprofit research institutes, retail and wholesale business, and services, which in reality are actively engaged in technology imports. In other words, the Bank of Japan figures tend to understate technology exports, whereas the Bureau of Statistics figures tend to understate technology imports.

(5) Fiscal year in Japan starts in April and ends in March of the following year.

(6) Total R&D expenditures including humanities and social science is 5246.2 billion yen in 1980. This means that approximately 89 percent is spent in natural science fields. This ratio has been rather stable throughout the period.

(7) About 25% of utility models is said to qualify as patents according to a qualitative judgement (Japan Institute of Invention and Innovation, 1978, p.27).

(8) In 1980, there were 3133 applications for patents in the Soviet Union, of which 3130 were by foreign nationals.

(9) Dummy variables employed include 1971 dummy (1971=1, 0 elsewhere) for Japan, 1972 dummy (1972=1, 0 elsewhere) for the United States, and 1974 dummy (1974 and after =1, 0 elsewhere) for the Soviet Union. The Soviet function has the following form: Y=f(D*X, T, D*T), where D is the dummy variable.

(10) Form A for the Soviet Union includes a dummy variable to take account of the trend shift which has been observed since around 1974. The parameter on the time trend variable reported here has been obtained by adding the two parameters which apply for the period since 1974. The parameter in the case of form B is for the entire period from 1965 to 1980.

(11) See Kravis (1951), Posner (1961), Keesing (1967).

(12) Ministry of International Trade and Industry, Management Analysis of the World's Corporations, an International Comparison, various years.

Reference:

Bureau of Statistics, Prime Minister's Office. Report on the Survey of Research and Development (J). Tokyo, various years.

Ishikawa, Sotaro. Guide to the Soviet Patent System (J). (New Edition). Japan Institute of Invention and Innovation. Tokyo. 1982

Japan Institute of Invention and Innovation. A Basic Study on the Trends in Industrial Activity and Patent Application (J). Tokyo, 1978.

Keesing, Donald B. "The Impact of Research and Development on United States Trade." Journal of Political Economy. Vol. 75, February 1967.

Kravis, Irving B. "'Availability' and Other Influences on the Commodity Composition of Trade." Journal of Political Economy. Vol. 64, April 1956.

Ministry of International Trade and Industry, Bureau of Industrial Technology. International Comparisons of Industrial Technology (J). Tokyo, 1982.

Posner, M.V. "International Trade and Technical Change." Oxford Economic Papers, Vol. 13, No. 3, October 1961.

Schmookler, Jacob. "Economic Sources of Inventive Activity." The Journal of Economic History. Vol. 22, No. 1, March 1962.

Science and Technology Agency. White Paper on Science and Technology (J). Tokyo, various issues.

Science and Technology Agency. Indicators of Science and Technology (J). Tokyo, various issues.

Usui, Mikoto. "International Transfer of Industrial Technology: --- An Appraisal of the Japanese Performances in Latin American NICS." Latin American Studies. No. 6, 1983.

R & D Managment Systems in Japanese Industry
H. Eto and K. Matsui (editors)
© Elsevier Science Publishers B.V. (North-Holland), 1984

Chapter VII BEHAVIOUR OF JAPANESE R & D ORGANIZATIONS --- CONFLICT AND ITS RESOLUTION BY INFORMAL MECHANISMS ---

HAJIME ETO

Institute of Socio-Technology
The University of Tsukuba
Sakura, Ibaraki 305, Japan

Herein are examined the prevailing hypotheses that Japan is a nation of consensus and that the performance of Japanese R&D is based on the strong national unity. The severe national conflict concerning R&D and its effect on national R&D organizations are described. The legally weak functions of R&D-concerned ministries and agencies are shown to be partly as a result of the conflict concerning the supreme policy. Various informal mechanisms which are devised to compensate for the weak functions of formal mechanisms are discussed. Then the ground for the informal mechanisms to work effectively is identified. Finally, comments are presented on the recent international criticism over Japanese technology behind her export surplus, and an idea behind Japanese technology is presented in contrast to the traditional one held by the Western culture.

1. PROBLEM STATEMENT

The economic success of Japan was at first explained by her clever adaptation to Western technology. This explanation is, at least partly, true by recalling the theory that Western science has started with the learning from Eastern one. As the economic success of Japan lasts long, a doubt is raised against the above explanation because the imitation can not afford to support the economic progress for so many years. A couple of attempts [Dobrov 1977, Straszak, Eto 1980A, Eto 1980B] have been made to explain her long-lasting economic success as a fruit of her proper innovation based on her own original technology system. These attempts are all based on the recognition that her proper management of innovation works with the aid of her strong national unity.

The cultural unity of Japan has historically been formed by her linguistic unity and her geographic as well as linguistic isolation from the rest of the world. Such a historically formed unity does not straightly apply to modern problems like technology. An example is her persistent conflict concerning the diplomatic policy. Her public opinion has been split between pro-Americanism and neutralism or between the rearmament and the disarmament for these three decades till this lack of consensus caused a severe conflict between Japan and USA. Meanwhile USA remained unified on its diplomatic policy till the late 1960's and the two major parties in UK and West Germany have not been so severely opposing each other on their diplomatic policies till recently. This fact makes it questionable to apply the national unity theory to the modern fields like technology in a straight manner.

The economic unity of Japan is sometimes called the Japan Inc. This owes greatly to governmental control of the financial market in the postwar period when the capital was extremely scarce. Even today the strategy and tactics of the banks on interest rate, the opening and closing of branch offices and so forth need the permission of the Ministry of Finance. Such a legally authorized control by the government does not exist in technology innovation. The co-operation of independent firms has been eased by the coordination by Zaikai (a circle of businesses and the community of top businessmen, like Keidan-ren and Nikkei-ren) in the economic fields, but only the specialized professional societies exist in the technological fields. Loans from the government-controlled special banks to private firms share larger portion of the whole loans in Japan than in the other industrialized countries. This helps the government to control the industry. But, oppositely, the governmental R&D expenditure shares much smaller portion of the whole R&D expenditure in Japan than in the others [Eto 1980A]. Hence financial control by the government for the coordination of private R&D activities is quite weak in Japan. These facts demand a careful examination of the assumption that the strong national unity of Japan greatly contributes to her R&D.

The Section 2 shows the existence of a strong conflict concerning science and technology policy and the lack of a unified or cohesive national policy on R&D. The Section 3 identifies what invisible mechanisms operate for national unification of the R&D strategies. The Section 4 identifies what mechanisms are taken for the coordination of R&D activities across independent private firms. The Section 5 discusses how the peculiar decision system of Japan makes governmental guidance of R&D activities acceptable or even wanted by the industry. Finally the Section 6 is devoted to a discussion on the recent international criticism over Japanese technology.

The methods of analyses herein are those used in political science, administrative science, management science, sociology and anthoropology. These disciplines have so far been exclusively concerned with diplomatic policy, financial policy, production management, personnel management, local communities, families, cultural aspects of human activities and so on. But herein are investigated the science and technology policy, scientific organizations and communities, innovation aspects of human activities which are left uninvestigated by these disciplines. The stresses are on the informal aspects of organizational activities because the difference between formal and informal structures is quite significant in Japan. Especially the operation in hierarchical control lines and staff functions outside the hierarchical lines are closely discussed because they both take the similar forms to the Western ones but work very differently. The investigation of these aspects is expected to clarify characteristics of the Japanese R&D strategies in contrast to the Western ones.

2. UNSYSTEMATIC DISTRIBUTION OF FORMAL POLICY FORMATION ORGANS

2.1 Administrative Structure in Science and Technology Field

The administrative structure in the science and technology fields in the Japanese public sector is outlined in Figure 2.1 and is more detailed in Figure 2.2. Figures 2.1 and 2.2 present the following characteristics.

(1) Distributed system: There is no single ministry nor agency for science and technology administration.

(2) Specialty division system: Each ministry (e.g., the Ministry of Transport) administers directly its own research institutes (e.g., Ship Research Institute and Port & Harbour Research Institute), indirectly supervises the research institutes (e.g., Railway Labour Science Research Institute) of the public corporations (e.g., Japan National Railway) in its administration field and more indirectly supervises the private research institutes (e.g., the research institutes of private railway companies) in the field.

(3) Overlapping system: The scope of an organ is also covered by that of another at least in name. The Science Council and the Council for Science & Technology enscope the same field in their names. Further, the Council for Science & Technology stands in parallel to more specialized councils (e.g., the Council for Ocean Development) instead of standing in hierarchy. (Here and hereafter "the Science Council" denotes "the Science Council of Japan" or JSC which is distinct from Science Council attached to the Ministry of Education).

The characteristics stated in (1) and (2) are not surprising but the characteristic stated in (3) may be surprising. The last characteristic reveals that the administrative structure for science and technology has not been designed with the cohesive strategy. This further indicates that the distributed system stated in (1) is not the intendedly preserved nor designed system but the result of the past conflicting systems. The subsection 2.2 describes the historical course in which these organs have been set up.

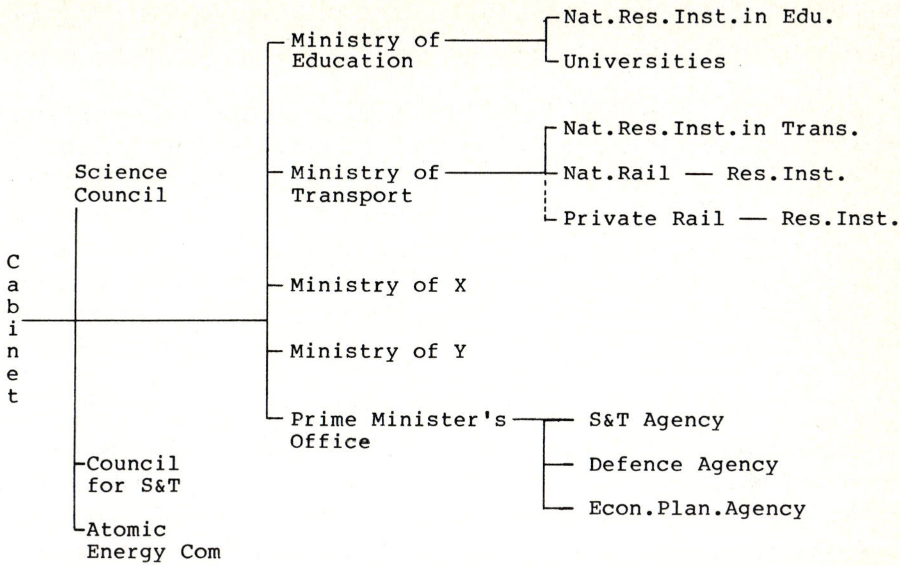

Figure 2.1 Outline of Administration Structure
(made from documents of the Administration Management Agency)

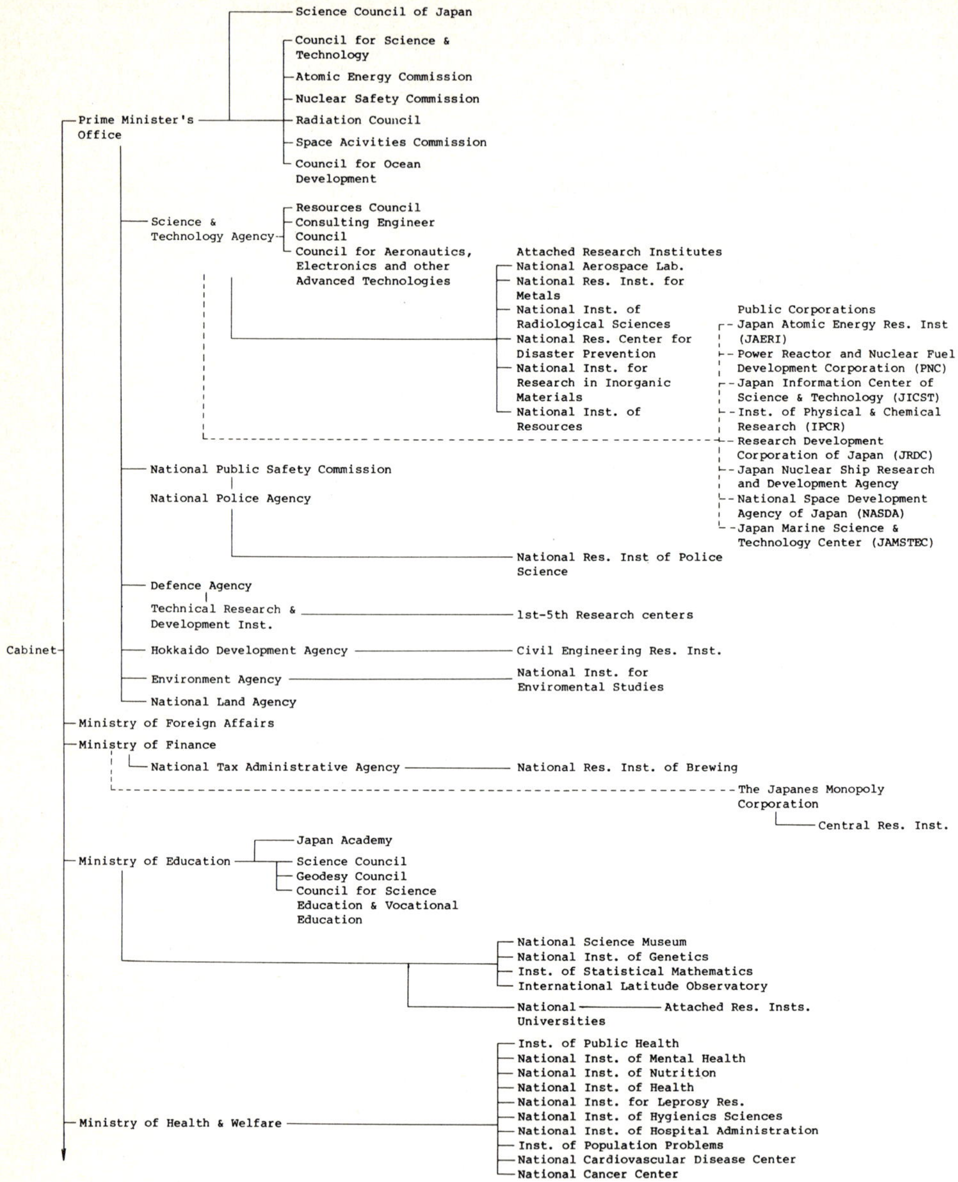

(Source: [STA sta, 1981])

Figure 2.2 Administrative Structure of Science and Technology in Public Sector

- Ministry of Agriculture, Forestry and Fisheries
  - Agriculture, Forestry and Fisheries Research Council
  - Attached Research Institutes
    - National Inst. of Agricultural Sciences
    - Central Agricultural Experiment Station
    - National Inst. of Animal Industry
    - National Grassland Res. Inst.
    - Fruit Tree Res. Station
    - Vegetable and Ornamental Crops Res. Station
    - National Res. Inst of Tea
    - National Res. Inst. of Agricultural Engineering
    - National Res. Inst. of Agricultural Economics
    - Sericultural Experiment Station
    - National Inst. of Animal Health
    - National Food Res. Inst.
    - Inst. for Plant Virus Res.
    - Tropical Agricultural Res. Center
  - Forestry Agency
    - Forestry and Forest Products Res. Inst.
  - Fisheries Agency
    - Regional Fisheries Res. Lab.
    - National Res. Inst. of Aquaculture
    - National Res. Inst. of Fisheries Engineering
  - - - Public Corporation
    - - - Inst. of Agricultural Machinery
- Ministry of International Trade & Industry
  - Industrial Technology Council
  - Agency of Natural Resources and Energy
  - Patent Office
  - Agency of Industrial Science & Technology
    - National Res. Lab. of Metrology
    - Mechanical Engineering Lab.
    - National Chemical Lboratory for Industry
    - Government Industrial Res. Inst. Osaka
    - Government Industrial Res. Inst. Nagoya
    - Fermentation Res. Inst.
    - Res. Inst. for Polymers and Textiles
    - Geological Survey of Japan
    - Electrotechnical Lab.
    - Industrial Products Res. Inst.
    - National Res. Inst. for Pollution and Resources
    - Government Industrial Development Lab., Hokkaido
    - Government Industrial Res. Inst. Kyushu
    - Government Industrial Res. Inst. Tohoku
    - Government Industrial Res. Inst. Shikoku
    - Government Industrial Res. Inst.. Chugoku
- Ministry of Transport
  - Council for Transport Technics
  - Ship Res. Inst.
  - Port & Harbour Res. Inst.
  - Electronic Navigation Res. Inst
  - Traffic Safety and Nuisance Res. Ins.
  - Meteorological Agency
    - Meteorological Res. Inst
  - - - Japanese National Railways
    - Railway Technical Res. Inst.
    - Railway Labour Science Res. Inst.
- Ministry of Posts & Telecommunications
  - Radio Technical Council
  - Radio Res. Lab.
  - - - Nippon Telegraph & Telephone Public Corporation
    - Telecommunication Lab.
  - - - Kokusai Denshin-Denwa Co., Ltd. (KDD) (Japn's Overseas Radio & Cable System)
    - Res. Lab.
  - - - Nippon Hoso Kyokai (NHK) (Japan Broadcasting Corporation)
    - NHK Technical Res. Labs.
- Ministry of Labour
  - Res. Inst of Industrial Safety
  - National Inst. of Industrial Health
- Ministry of Construction
  - Geographical Survey Inst.
  - Public Works Res. Inst.
  - Building Res. Inst
- Ministry of Home Affairs
  - Fire Defence Agency
    - Fire Res. Inst.

2.2 Historical Background of Administrative Structure

The distributed system described in (1) in the subsection 2.1 is explained as a result of the historical situation. When the Japanese constitutional system started in 19c., there was no need nor real ground to promote science and technology. Hence the administration of science and technology was naturally distributed among various ministries. For example, the Ministry of Education was concerned with science education and the university research budget, the Ministry of Welfare was concerned with medical research outside universities which were supervised by the Ministry of Education, and so on. In the war period (around 1935 - 1945) the Ministries of Army and Navy were the most influential over the R&D policy. After the Ministries of Army and Navy were disorganized in the postwar period, a democratization and peace-seeking movement of scientists started as a reaction to the previous government-controlled and defence-oriented R&D policies.

The democratization and peace-seeking movement of scientists arrived at establishment of the Science Council as their own unified and independent organ. The establishment was in the beginning of 1949 with the support of the socialist cabinet (1947 -48) and the socialist-coalition cabinet (1948 - 49) [Science Council]. The Act for Establishment of the Science Council which passed the Diet in 1948 wherein the Socialist Party was ruling declares that the Science Council represents the all Japanese scientists to the domestic and abroad. This was understood to imply that the Science Council is the supreme organ in science.

The Japanese word "science" denotes what is called "Wissenschaft" in German and enscopes humanities, social sciences, natural sciences, engineering, agriculture and medicine. So does the Science Council.

When the Science Council was established, the cold war had already started and the Korean war followed it. The Liberal Party which took the power from the socialist-coalition government a few days after the establishment of the Science Council proposed an idea of establishing a new agency and a new council of science only four years after the establishment of the Science Council. After some political debate, this idea was realized. Science & Technology

Agency was established in the Prime Minister's office in 1956. Council for Science & Technology was established in 1959 as attached to the Prime Minister's Office. The overlapping of Science Council and Council for Science & Technology is a result of the political conflict.

Prior to the establishment of Council for Science & Technology, Atomic Energy Commission and Radiation Council had already been established. As a result of a political compromise with the opposing power, the newly established Council for Science & Technology failed to stand above the already established commission and council. Following this example, the commissions and councils established afterwards stood in parallel to Council for Science & Technology. This resulted in a complicated overlapping of the organs. The establishment of the commissions and councils after Council for Science & Technology does not directly reflect the political conflict. This rather reflects evolution of new scientific fields beyond the scope which the Science Council and Council for Science & Technology were intended to cover.

The Science Council consists of voters and council members elected by the voters. The number of voters exceeds 200,000 in 1982 [Science Council]. Qualifications of the voter are quite modern as reflecting the democratic atmosphere at the time of establishment: The most important criteria are publication and research experience rather than the traditional criteria of title, educational background and occupation. Hence the 200,000 voters include highschool teachers and highschool-graduate engineers in private firms who have published their works. Each voter is registered to one of the seven divisions. The problem is this division system. The divisions are humanities; law and political sciences; economics; natural sciences; engineering; agriculture and medicine. These seven reflects the faculty structure of the established universitieS in the 1940's but does not necessarily reflect the structure of the modern institutes today. Each division has an equal number (30) of the council members, totally 210 (= 30 $\times$ 7) council members. This structure has remained unchanged in these three decades since then. This equality system among divisions was quite renovative at that time as opposed to even the reformed structure of the Japan Academy in the postwar time wherein humanity (30) and pure science (31) dominate the fields of law (24), economics (10), engineering (17), agriculture (12) and medicine (20). Since then, however, the engineering faculties grew very quickly and private research institutes employed a great number of engineers. Hence the voters and the council members in the engineering division began to feel a gap between the equality structure and the real situation. Another severe problem is the evolution of interdisciplinary fields across the divisions.

Table 2.1 Constituency System of the Science Council

Div.	nation-wide			local	total
	subfield		field		
1st	Literature	3	11	7	30
	Philosophy	3			
	Pedagogy, Psychology, Sociology	3			
	History	3			
2nd	Law in general	2	13	7	30
	Public law	2			
	Civil law	2			
	Criminal law	2			
	Political sci	2			
3rd	Economics	5	13	7	30
	Commerce, Business	5			
4th	Mathematics	1	12	7	30
	Astronomy	1			
	Geophysics	1			
	Physics	1			
	Chemistry	1			
	Zoology	1			
	Botany	1			
	Geology	1			
	Mineralogy	1			
	Geography	1			
	Anthropology	1			
5th	Appl. Physics	1	14	7	30
	Mechanical Eng.	1			
	Electrical Eng.	1			
	Ship-building	1			
	Civil Eng.	1			
	Mining	1			
	Metallurgy	1			
	Appl. Chemistry	1			
	Architecture	1			

to be continued

Div.	nation-wide		local	total
	subfield	field		
6th	Agriculture 1 Agr. Chemistry 1 Forestry 1 Fishery 1 Agr. Economics 1 Agr. Eng. 1 Zootechny 1 Sericulture 1	15	7	30
7th	Fundamental Med 4 Clinical Med 4 Public Health 2 Dentistry 2 Pharmacology 3	8	7	30

(Source: [Science Council])

The Science Council and its supporters were the first group in Japan who recognized the importance of the evolving interdisciplinary fields. In fact, their predecessors formed interdisciplinary research teams in the 1930's and won the various patents and the Nobel prizes. The key members of the Science Council were (and also are) from these teams. For another example, biochemistry grew with the great help of the Science Council and the same applies to plasma physics and technology. Table 2.2 indicates the sincere interest of the Science Council in the evolving interdisciplinary fields. Despite this, the Science Council failed in taking the initiative for undertaking these evolving fields. This may be explained as an effect of the zero-growth of the financing for the Science Council.

Table 2.2 Institutes and Research Facilities Recommended by the Science Council

Established		Not yet established	
Fundamental physics	1950	Atmospheric physics	1965
Nuclear study	1953	Earth Sci	1965
Astrophysical observatory	1953	Information eng & sci	1965
Solid state physics	1956	Crystallography	1967
Protein research	1956	Regional geography	1967
Ocean research	1958	Human behaviour	1967
Mathematical sci	1958	Breeding sci	1967
Radiological sci	1958	Environmental radioactivity	1968

to be continued

Established		Not yet established	
National archive	1959	Radiation effects	1968
Plasma physics	1959	Organic chem	1968
Polar research	1961	Biophysics	1968
Language & cultures of Asia & Africa	1961	Structural eng	1968
Research reactor	1962	Social material center	1970
Space & aeronautical sci	1962	Biologically active natural products	1971
Cosmic rays	1962	Complex chem	1971
Nuclear physics	1962	Quaternary	1971
Univ. computer centers	1963	Composite materials	1971
Primate research	1964	After-effects of A-bomb	1971
Welding	1964	Biotron	1971
Ethnological study	1965	Polyphase flow	1973
Molecular sci	1965	System sci	1974
Biology	1966	Human genetics	1974
Japanese literature	1966	Basic bio-eng	1975
Physiology	1967	Echology	1977
Radio-telescope	1971	Energy eng	1977
Water resource	1971	Mineral resource	1977
Photon factory	1974	Europe language	1977
		Fundamental Math	1979

Year of establishment (left column), Year of recommendation (right column)

(Source: [Science Council])

Since the establishment of the Science Council and its conflict with the government, the government declined to increase the budget for the Science Council despite the inflation and the economic growth [Science Council]. The tight finances of the Science Council made it impossible to expand its activities to the evolving fields without the cutting of the budget for the existing activities. The cutting of the existing activities was difficult and therefore the expansion to the evolving fields was sacrificed. This made the scientists in these fields disappointed with the Science Council and they began to ask for the budget directly from the government. Upon this groundwork the government established the commissions and councils for the interdisciplinary fields in the 1960's and 1970's.

Council for Science & Technology which was established for the purpose of adapting of science and technology to the economic growth could not serve this purpose in an effective way. For the authorization of the newly established council, the senior scientists and economists who experienced high positions like deans or presidents of universities or companies were selected as council members.

This made the activities of Council for Science & Technology quite inadequate for the evolving fields. This situation forced the additional commissions and councils to be set up independently of Council for Science & Technology.

The complicated overlapping of the various organs for science and technology is explained only as a result of the political conflict concerning science and technology policy. This historical fact contradicts the general feeling that Japan is a nation of consensus in every field.

For the detailed information of the idea underlying the overlapping, the law, act or cabinet order for the establishment of the commissions and councils are useful and will be described below.

The statements in the Act of the Science Council which passed the Diet in 1948 imply that the Science Council is the only organ of this kind and that it can directly communicate to the government. Contradictingly, the Act for Establishment of Council for Science & Technology states that it acts as the liaison between the Science Council and the government. Thus the Science Council has lost its direct channel to the government since then. This fact gives the impression that Council for Science & Technology stands above the Science Council. Indeed the booklets of Science & Technology Agency [e.g., Sta sta] say that Council for Science & Technology is the supreme council on science and technology. The inconsistency of the two Acts is a result of the political conflict and compromise, reflecting the lack of national consensus on science and technology.

The Act for Establishment of Atomic Energy Commission which passed the Diet in 1955 defines its purpose as the democratic operation of atomic energy administration. Contrastingly, the Act for Establishment of Space Activities Commission which passed the Diet in 1968 defines its primal purpose as the comprehensive and systematic promotion of the national space activities and its secondary purpose as the democratic operation of the administration of the national space activities. The time passage of 13 years has shifted the democratic operation of the administrations from the primal purpose to the secondary one and pushed up the comprehensive and systematic promotion to the primal one. The Act for Establishment of Council for Science & Technology in 1959 defines its purpose as the promotion of science and tecnnology while the Act for Establishment of Space Activities Commission defines its primal purpose as the comprehensive and systematic promotion. The time passage of 9 years augments the promotion with the adjectives of comprehensive and systematic.

The Act for Establishment of Atomic Energy Commission (this part passed the Diet in 1955) and Nuclear Safety Commission (this part was augmented by the Diet in 1978) defines their authorities as planning, deliberating and deciding the policies on the related affairs. After 1955 the authority of decision is eliminated from the law, acts and cabinet orders of the commissions and councils except the Act for Establishment of Space Activities Commission in 1968 which allows authority of decision to the Commission.

The primal purpose and authority of each of these commissions and councils are to deliberat and advise policies of the related affairs to the government or the Prime Minister.

Table 2.3 Councils and Commissions

Year	Organ	Purpose	Scope	Advisory for
1949	Sci.Counc.	Unity of scientists	Sci.Tech.Hum. Med.	Government
1956	Atomic E.Com.	Democratic Admin.	Atmoic Energy	Selfdecision
1978	Nuclear Safety Com.	Democratic Admin.	Nuclear Safety	Selfdecision
1957	Radiation Counc.	Standardi-zation	Radiation	Concerned Ministers
1959	Counc. S&T	Promotion of S&T	Nat.Sci.& Tech	Prime Minister
1968	Sapce A. Com.	Comprehen-sive Prom.& Democ.Adm.	Space Activity	Prime Minister
1971	Ocean D.Counc.	Basic & Compreh.	Ocean Develop.	Concerned Ministers

(Made from related laws and acts)

2.3 Dominance of Politics over R&D in Formal Organizations

The subsection 2.2 described the historical development of formal organs for science and technology wherein the political motivation was dominant over the promotion of science and technology. It may be natural that the government did not take R&D itself as important so much when Japanese science was underdeveloped. Today, however, R&D is highly emphasized in the Japanese society. Hence it may be expected that the government has now begun to handle the R&D policy with no or little interest in political or ideological issues.

The Prime Minister's Office started the harsh criticism against the Science Council in 1981 and proposed its reform in 1982. Actually the proposed reform plan is aimed not at promoting the evolving interdisciplinary fields across the existing professional societies nor at mobilizing the R&D resources (e.g., researchers) presently allocated outside the main stream for the evolving fields. Oppositely, it is aimed at replacing the direct election system of the council members by the great number of voters with the indirect recommendation system by the existing professional societies to the Prime Minister and the nomination of council members by the Prime Minister. This is equivalent to replacing the present election district of major size according to the university faculty structure with the segmented election district of minor size according to the specialized scope of existing professional societies. Such an indirect election system is generally favourable to existing big powers and unfavourable to evolving small groups the later of which are usually more important than the former in R&D. Hence it is quite doubtful that the government is really attempting to modernize the Science Council. The press reports that the government hates the election system. Small interdisciplinary societies are now worried about with which societies to make coalitions in recommendation of council members.

The Japanese education system is quite centralized and is under the strong control of the Ministry of Education. The textbooks at the primary and secondary education levels are all supervised by the Ministry of Education both in the public and the private schools. Personnel management of teachers in public schools at the primary and secondary levels is legally left to local authorities but the latters are under the financial control of the central government (the so-called 30% autonomy and 70% centralization). Even for private universities, the setting-up of new departments and graduate schools must be authorized by the Ministry of Education and almost all private universities are financially subsidized by the Ministry because the tax system discourages financial contribution of individual persons and private organizations to private universities. Needless to mention the financial control of national universities by the Ministry. This implies that the Ministry of Education is responsible for the science education and R&D activities in universities as well as primary and secondary schools irrespective of public or private ones.

The Ministry of Education has been more interested in the education of patriotism than the education of science and mathematics since the 1950's. The school teachers who neglect modern teaching methods of science using modern experiment equipment and the modern topics like electronics, the probability theory and statistics are not punished, but the teachers who are reluctant to the patriotic education are often severely mistreated. The policy to promote the patriotism through history textbooks supervised by the government caused a severe conflict with other Asian nations in 1982. In contrast to such earnestness in the patriotic education, the Ministry of Education have never accused teachers for the fact that most of the highschool students have remained untaught of the probability theory and statistics which have been stipulated to be taught by the Ministry. Further, the Ministry of Education recently relaxed the "burden" of science and mathematics requirement in schools despite the demand of industry.

The reason of not teaching these modern topics is that the elderly teachers did not study these topics at their training. Hence the solution is simply to retrain teachers for these modern topics as the world-widely known training or retraining program in private firms. But the Ministry is not earnest enough for the retraining program. This situation forced private firms to teach their workers the elementary probability theory and statistics for quality control activities in the firms. As its result, the knowledge of statistics and the probability theory among the Japanese workers is the world top [Rieker & Sullivan]. If private firms can do, why can not the Ministry? The answer is that private firms are more interested in science education than ideology education while the

Ministry has the opposite view. This may be the tradition of the Ministry of Education. The Ministries of Army and Navy were seriously interested in the promotion of science education in schools in the 1930's. This demand was reflected in the revised school textbooks, and relatively advanced topics were stipulated to be taught in schools. But actually, patriotic education was more emphasized till 1945. One of the reasons may be that science education is more costly. Now science training is left to firms which employ secondary school graduates. The firms are forced to pay what the government should pay.

The majority of politicians (at least, of the present government party) are quite pragmatic and reflect the interest of industry which is more concerned with innovation than ideology. But some politicians are ideologists rather than pragmatists. While the majority are busy with industrial policy making, the minority concentrate on the education policy and form "the family of the education congressmen" which is their nickname given by the press. Actually politicians of this "family" formed the Education Committee in the government party which is naturally influential over education policy. The recognition that Japan is tightly unified around industry is not necessarily true for science and technology.

Science organizations and science education have been distorted for the political reason. This reflects the lack of national consensus on science and technology.

2.4 Control Function of Science & Technology Agency

Science & Technology Agency (STA) belongs to the Prime Minister's Office (PMO). The character of PMO is quite ambiguous and no political scientist seems to have been successful in giving a clear idea about it. Officials working in PMO sometimes claim that PMO is the General Staff Office of the government. This claim is supported only in that some of the agencies in PMO have no direct control power on real activity. But the function of these agencies is very different from that of the staff to the president of the US government or American firms and even from the Japanese staff in the General Staff Office of the Army and Navy. First of all, the number of the staff in PMO is enormous. STA which is one of the nine diversified agencies in PMO has 2,171 regular personnel including secretaries and researchers in attached research institutes in the 1981 fiscal year. Its intra-agency office (head office) alone has 486 personnel including secretaries in that fiscal year. Among the agencies in PMO is the Defence Agency which is the counterpart of the Defence Ministry of other nations rather than the General Staff Offices. The same with some other agencies: National Police Agency; Environment Agency and National Land Agency, for example. These are the immatured agencies waiting for promotion to the title of ministry rather than staff offices. In fact, some ministries were once called the agencies and belonged to PMO.

- intra
  - Secretariat
  - Bureaus (Decoration; Personnel; Pension; Statistics)
- extra
  - Agencies (Imperial Household; Administration Management; Hokkaido Development; Defence; Economic Planning; Science & Technology; Environment; Okinawa Development; National Land)
  - Commissions (Anti-trust; National Public Safety; Public Nuisance)
- Councils, Commissions, etc. (40; including Council for Science & Technology)
- Training Institute, etc. (3)
- Committees, etc. (3, including the Science Council)

(July 1, 1982)

Figure 2.3 Administrative Structure of the Prime Minister's Office

(made from documents of the Administration Management Agency)

The Act for Establishment of Science & Technology Agency defines its function as a coordination organ between concerned ministries on science and technology except for university research. This implies that STA is authorized not to form the strategy and tactics but to coordinate the strategies and tactics formed by other ministries. The problem is what is the coordination power of STA in the situation that the other ministries have historically owned strong powers. If the activity of STA is based on the strong direction of the Prime Minister in that the macroscopic strategy is first directed by the Prime Minister and then the microscopic tactics is coordinated by STA, the coordination is unregistedly executed. But the bottom-up decision system of Japan and the weak interest of top politicians in science and technology lead to situations wherein each minister is not forced to follow the coordination of STA. STA is relatively young and was established in 1959 when the other ministries had already their own research institutes which are still now administered by these ministries.

The situation that other ministries have stronger power than STA requires STA to make a highly attractive plan. For this purpose, STA will need special information-creating organs within it. STA has six research institutes, laboratory and center and supervises four research institutes and centers (altogether ten) which might serve for this purpose of information creation. In fact the characteristic structure of STA regarding the administration of research institutes and centers facilitates for the research institutes and centers to serve for this purpose. The Act for Establishment of STA defines its research institutes and centers to be administered by the respective bureaus. This structure facilitates for each institute, laboratory or center to assist its respective bureau in the policy formation in the respective fields. For example, Research Coordination Bureau includes Space Activities Planning Division and administers National Aerospace Laboratory. This structure makes it easy for the Bureau to be assisted by the Laboratory in forming space policy.

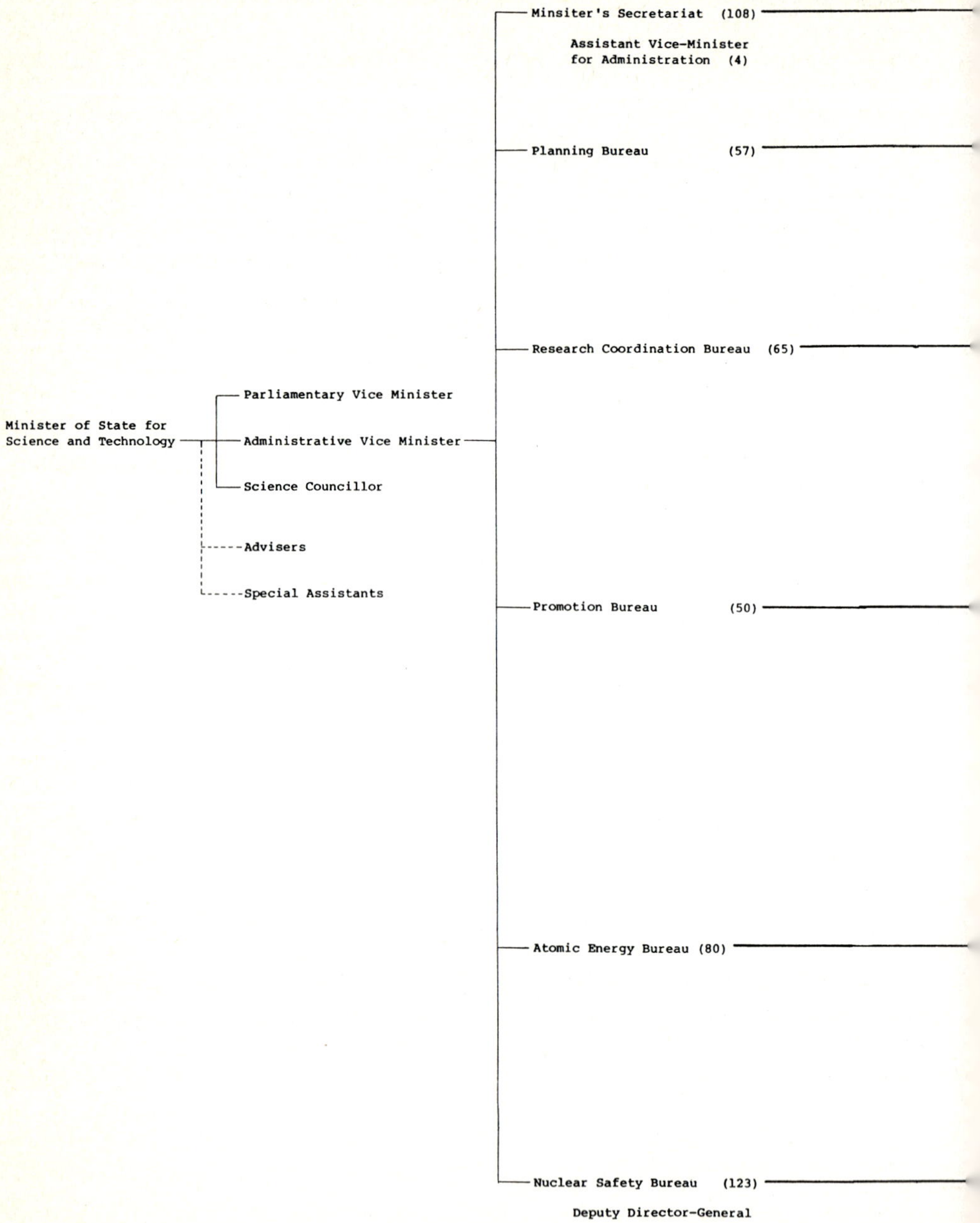

(Source" [STA sta., 1981])

Fig. 2.4 Administrative Structure of STA

- Personnel Division
- General Affairs Division
- Finance Division
- Science Counsellor (1)

Attached Research Institutes and Public Corporations — Advisory Organs

- Planning Division ---------- Council for Science and Technology
- Research Division
- Resources Division ---------- Resources Council
  - National Institute of Resources (42)
- Science Counsellors (5)

- Coordination Division ---------- Council for Aeronautics, Electronics and other Advanced Technologies
- Inter-ministerial R&D Div. — National Research Center for Disaster Prevention (118)
- Life Technology Division ---------- Space Activities Commission
  - National Aerospace Laboratory (475)
- Space Activities Planning Division
- International Space-Affairs Division — National Space Development Agency of Japan (903)
- Space Development Division — National Space Development Agency of Japan
- Ocean Development Division ---------- Council for Ocean Development
  - Japan Marine Science and Technology Center (126)

- Promotion Division ---------- Consulting Engineer Council
- Public Relations Division
- International Affairs Div.
- Invention Division
- Institutes Administration Division
  - National Research Institute for Metals (464)
  - National Institute for Research in Inorganic Materials (168)
  - Institute of Physical and Chemical Research (623)
  - Research Development Corporation of Japan (73)
  - Japan Information Center of Science and Technology (332)

- Policy Division ---------- Atomic Energy Commission
  - Mito Atomic Energy Office (10)
- Research and International Affairs Division
- Power Reactor Development Division — Power Reactor and Nuclear Fuel Development Corporation (2,681)
- Technology Development Div. — National Institute of Radiological Sciences (416)
- Nuclear Fuel Division
- Institutes Administration Division
  - Japan Atomic Energy Research Institute (2,396)
  - Japan Nuclear Ship Research and Development Agency (143)

- Nuclear Safety Division ---------- Nuclear Safety Commission
- Reactor Regulation Division
- Nuclear Materials Regulation Division
- Safeguards Division
- Radiation Protection Division ---------- Radiation Council

STA owns four research institutes, a reseaıch laboratory and a research center and supervises two research institutes and two centers. Among these ten organs, three are in the big-science fields (atomic energy, aerospace and ocean) and one is the science and technology information center (JICST). This fact may lead to the expectation that STA is informationally in the eminent position especially in the big-science fields if these research organs have the high performance. But these are relatively small in size for the big fields. National Aerospace Laboratory has 472 staffs, Marine Scinece & Technology Center has 133 staffs and JICST has 330 staffs including non-researchers like secretaries in the 1982 fiscal year. Atomic Energy Research Institute has 2,404 staff including non-researchers in the 1982 fiscal year but this number is not big enough for this field. These small personnel numbers indicates that the information-creation function of STA is quite limited.

STA is allocated over 40% of the special coordination funds for promoting science and technology every year (166 out of 392 billion yen in the 1982 fiscal year). This is the only means of budgetary control at the hand of STA. But it amounts to only one-third of the budget for R&D in national universities (438 billion yen in the 1979 fiscal year) which is under control of the Ministry of Education.

As the establishment of new departments or faculties in national universities (and also private universities through the inspection and the financial aid) is under control of the Ministry of Education, the Ministry could control the promotion of evolving interdisciplinary fields through the establishment of new departments or faculties. But the action of the Ministry has a long timelag after the demand from the within-university [Tsukahara & Yamada]. The Ministry waits till the unanimous agreement is attained about it within university. The only field in which the Ministry does not take this consensus-waiting strategy is the social science. The Ministry pushes the modernization of social science faculties despite resistance from faculty members. Many social science professors are pro-Marxists after the nationalistic-minded professors have left universities in the postwar period. The Ministry devotes significant effort to its reform long before within-university consensus is attained. The fact that the large amount of R&D budget is controlled by the Ministry of Education which executes its initiative mainly for the political or ideological affairs indicates the limit of the budgetary power of STA. Among the science-concerned ministries and agencies the Ministry of Education alone has the strong power often with the aid of armed forces of police to crash the teachers union but, inconsistently, this power has never been executed to the scientifically conservative university research. STA is forced to act in the situation that the university research and education remain out of date.

2.5 Staff Functions in Government

It is sometimes said that the staff function in the narrow sense of the word have never existed in Japan except in the army and navy [e.g., Tsuji]. The decision assisted by the relatively small number of staff is located in middle between the monarchical dictator decision and the democratic or participatry decision. The staff or advisors are needed, for example; when the top-down decision system is politically chosen but the top technologically lacks expertise; or when the democratic or participatry decision system is politically chosen but there is no time for deliberation by a great number of participants. In the course of her historical development Japan has almost always been in neither situation.

Her geographic isolation from the other powers has not demanded the quick decision making in the Japanese history except a few cases. This historical situation has allowed Japan collective decision by the slow deliberation. This decision system has in turn allowed the participation of relatively great number of high officials as compared to the decision system chosen under the situation of severe struggle with the neighbours.

The historical fact that the Emperor system has lasted about fifteen hundreds years is associated with the collective or bottom-up decision system which liberates the Emperor from the political responsibility. When political structure revolves, a former political leader who is responsible for the former leadership is evicted from the power. But the Emperor system has survived beyond the revolutions in Japanese history because the Emperor has remained free from the political responsibility under the cllective or bottom-up decision system.

The top is given only legal or nominal power in most of Japanese organizations. He seldom executes the power himself and usually leaves it to his men. In such a decision system all administrative officials are the staff to assist the decision of the top and at the same time are executives under the top. This situation makes special position of staff unneccessary.

The "ringi" system inforces the bottom-up decision system. The word ringi originally came from the written ringi petition [Tsuji]. Ringi means that a bottom officer proposes a plan in the ringi paper and circulates it to his boss and directors of the concerned sections. It can circulate vertically as well as horizontally in the hierarchical organization. If all of the directors of the concerned sections and his boss endorse it, it becomes the authorized decision and is kept as the authorized document [Tsuji]. The only place where the ringi system did not exist in Japan was the army and navy where quick decisions were needed and staff functions existed [Tsuji]. It may be concluded that the staff function is inconsistent to the ringi system in practice.

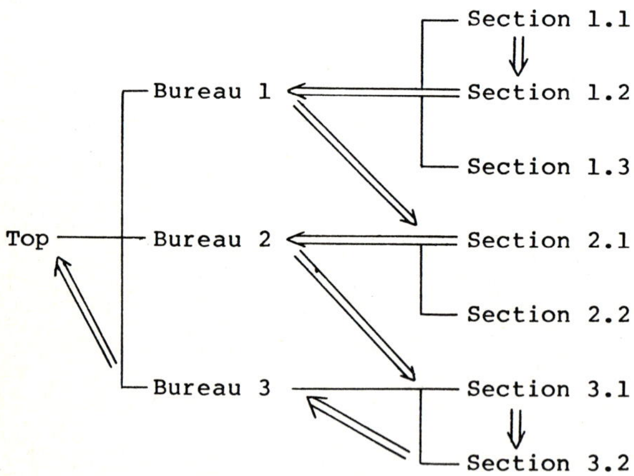

Figure 2.5 Circulation of Ringi

Different from the modern Western system of separation of the power among the legislative, administrative and judicial powers, the power is concentrated, at least nominally, on the top in many cultures. At the same time the various measurers are devised to prevent the individual arbitrariness. The universal principle based on religion, moral rule or ideology is one example of the measures and the collective decision is another. Under the condition of underdevelopment of the power separation system, the collective decision system enforced with the ringi system has been developed in Japan. This has prevented the training of staff as highly specialized professionals.

USA occupied Japan in 1945 and tried to reform the Japanese administrative system. An American study group discovered that the major difference between the American and Japanese administrative behaviour is the attitude of the officials to the specialist. According to the study, the American officials were (also are) relatively inclined to the position of specialists but the Japanese declined to it [Ito]. The Japanese officials claimed (also claim) themselves as generalists rather than specialists, but the American study group called them departmentalists in that their view was (also is) limited to matters within their own departments (ministries). Such a narrow-sighted view is quite inadequate for staff activities which require oppositely the broad-sighted view.

As one of the measures to break the closedness of departmentalists, the committee system composed of the experts from outside the bureaucratic organizations was strongly recommended by the American advisors. This system survives still now as private advisory boards to each minister as well as official advisory boards to each ministry. The boards are composed of experts mainly from the academic community and the press. They were expected to advise the administration and to revise bureaucratic behaviour in an effecive manner.

The idea of the private advisory board is quite old in Japanese history. The Confucian idea of benevolent administration encouraged the participation of Confucian experts in decision making as advisors to the top. The Imperial Constitution (1889 -1947) established the Privy Council as the supreme advisory board to the Emperor with the Councilmen being the experts such as experienced statesmen. The Senators of the Upper House which stood above the Lower House in the Imperial Constitution were also composed of experts as well as the Noblemen; the experts being the leading professors, experienced lawyers, retired high government officials and so forth. In the modernization process in the 1930's, the Cabint Council System was promulgated in 1935 though it was abolished in 1936. The Cabinet of Duke Konoe (1940 - 1941) highly evaluated the role of relatively young staff. These were intended to break the sluggish bureaucracy.

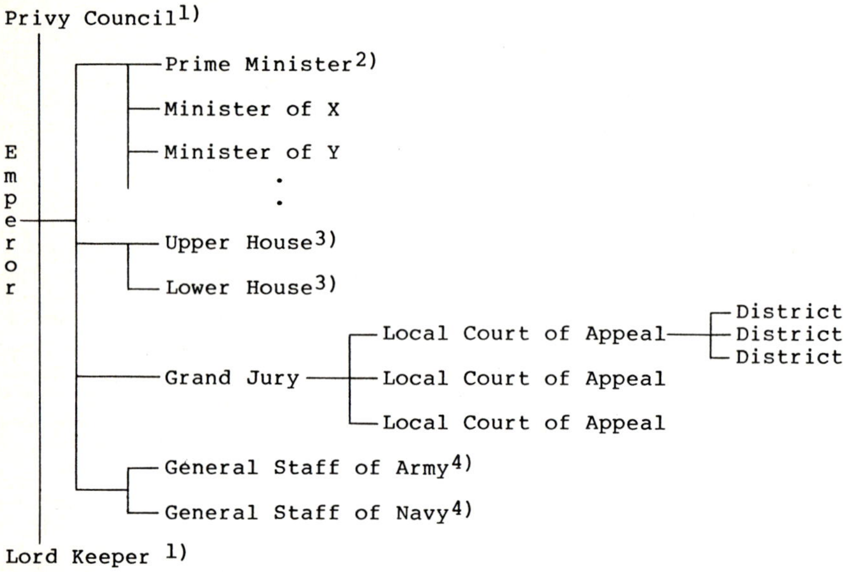

1) Now discontinued

2) P.M. was equal to the other minsters under Emperor

3) Upper House stood above Lower House (now the opposite)

4) Command of military operation was directly under Emperor and was independent of Cabinet (now under P.M.)

Figure 2.6 Imperial Constitution (1889 - 1947)

(made from the Imperial Constitution)

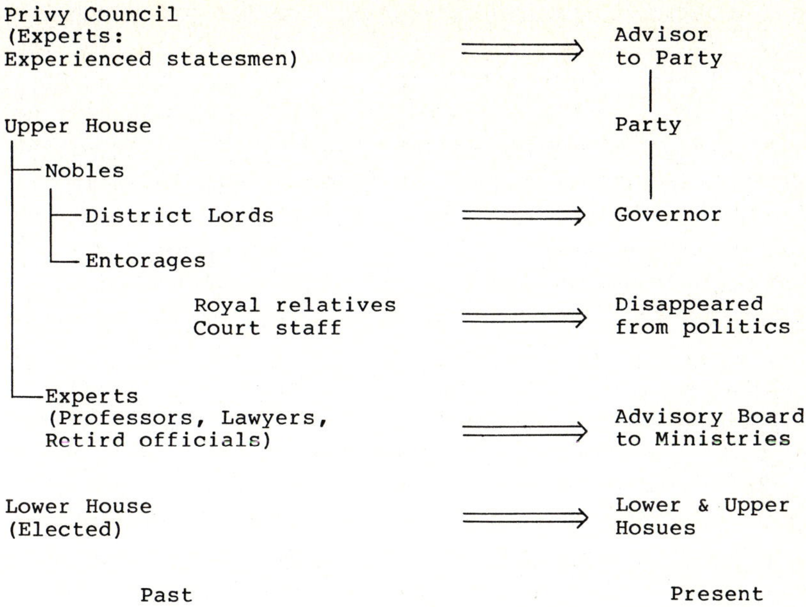

Figure 2.7 Transition of Deliberative Staffs

The initial intention of the advisory board to break the bureaucracy has been weakend in the course of time. Today most members of the advisory boards are senior or elderly experts selected by the bureaucratic decision in each related ministry or agency. In most cases their function is merely to ratify proposals by bureaucratic offices instead of proposing alternatives to stimulate the routine work in the bureaucracy. They play the role of liaison channels between the administration and the academic community or specialists communities, contributing to consensus formation between those worlds.

3. INFORMATION CREATION FUNCTION OF SCIENCE & TECHNOLOGY AGENCY

3.1 Publication Activities of STA

Science & Technology Agency (STA) which is the newest among the science and technology-concerned ministries and agencies does not administer the traditional fields which had already been administered by other ministries at the time of establishment of STA. Instead, STA administers the evolving interdisciplinary fields like space and ocean. In fact, the biggest research organs in these fields are administered by STA: National Aerospace laboratory (NAL) which is attached to STA and Japan Marine Science & Technology Center (JAMSTEC) which is supervised by STA. Other big fields administered by STA are atomic energy and science & technology information service: Japan Atomic Energy Research Institute (JAERI) and Japan Information Center of Science & Technology (JICST) both of which are supervised by STA and are the biggest research organs in the fields in Japan. If whole science and technology of Japan is centered around these fields, STA can have the strong initiative in whole science and technology of Japan.

The share of big sciences in whole science and technology of Japan was quite low as compared to the other nations till recently. Table 3.1 shows that the R&D expenditure of Japan is not concentrated to big sciences. In such a structure the position of STA is not a leading one. As STA lacks strongly organized ground in the existing frame of the Japanese society, STA needs a measure which does not require much assistance of existing organizations. The measure taken by STA is the publication activity appealing directly to the public.

STA publishes annually the White Paper on Science & Technology. The White Paper had attracted little attention till the 1970's. Recently, however, the Cabinet recognized its significant influence over the public. In fact the publication of the 1982 White Paper was suspended several months by the Cabinet. The White Paper on Economy which is naturally influential in such an economism nation of Japan was once suspended its publication for several days some

Table 3.1 Share of Public Fund for Big Sciences in the Whole R&D Expediture (%)

	Nuclear energy		Space		Ocean	
	1973	1979	1973	1979	1973	1979
Japan	3.20	4.86	1.85	2.44	0.55	0.99
France	13.01	8.51	4.65	3.59	0.88	0.51
FRG	4.83	4.83	2.92	2.02	0.66	0.46
UK	4.50	3.64	2.20	1.59	0.60	-
USA	4.97	4.30	10.45	7.73	2.06	2.07

(Source: [STA i, 1982])

years ago. This kind of suspension is quite rare and limited to the affairs of political importance in the Japanese administration wherein the function of the Cabinet is usually ceremonial authorization of the draft by the bureaucracy. This event indicates the increasing recognition of its importance.

Besides the White Paper on Science & Technology and the White Paper on Atomic Energy, STA publishes a series of books on the status quo and prospects in promising fields: life science; materials technology; ocean development; systems technology; electronics and communications; science and technology information systems. As far as the number of volumes is concerned, STA places priority on materials technology, ocean development and life science in this order. These books are edited by STA but the majority of editorial boards and many writers are from outside like universities, public or private research institutes and industry. A side-effect of this publication activity is to grow channels to the industry through organizing editors and writers therefrom. Historically STA has had poor channels to the industry but the number of editors and writers from the industry is growing. It will add to the asset of STA.

3.2 Delphi Forecasting Information Organized by STA

One of the most influential informations provided by STA may be the Delphi technology forecasting study which is made every five years

(1970-71, 1976, 1981-82). Its result is widely reported through the newsmedia and therefore is influential in target-setting and budgeting of R&D activities in the public and the private sectors. More specifically, its result is quoted at meetings of middle managers in industry for collective decision on target-setting and budgeting. This quotation is often unresistible to the members of the meeting and tends to lead the entire discussion of the meeting. In fact the latest Delphi study by STA (1982) was intentionally published at the time of budgeting for the 1983 fiscal year with the expectation of providing information of priority of R&D themes. Such popularity among, and an influential power over, managers may be due to its peculiar structure of the Delphi respondents.

The procedure of the STA Delphi study may be described as follows.

Step 0. The Planning Bureau (PB) is in charge of technology forecasting (TF).

Step 1. PB forms an ad hoc committee of TF study composed of experts (16 experts in the 1982 study) and specialized subcommittees (11 subcommittees in the 1982 study). Each subcommittee has the chairman from the parent committee and some additional members of respective expertise. (Table 3.2)

Step 2. Advised by the Committee and the subcommittees, (Step 2A) PB makes a questionnaire for each specialized field, (Step 2B) PB selects experts in each field as the respondents.

Step 3. With the consent of the Committee, PB announces the study will terminate with the second round.

Step 4. PB delivers the questionnaire to the respondents in each field and collects them (the first round).

Step 5. PB feeds back the distribution of the responses with some comments of the respondents and delivers the same questionnaire to the same respondents again (the second round).

Step 6. PB terminates the process at the second round as was announced.

Table 3.2 Structure of TF Committee in 1982

	Total	Univ.	Ind.	Journalism
Parent Com.	16	3	1	4
Subcom.	70(11)	20(3)	14(1)	2
1. Energy, Res.	7(1)	2(1)	2	0
2. Agr. Fish.	7(1)	1	0	0
3. Education	7(1)	2	0	2
4. Environment	6(1)	1	2	0
5. Med. Life	7(1)	5(1)	1	0
6. Urban	7(1)	1	2	0
7. Transport	5(1)	1	1	0
8. Information	5(1)	0	3	0
9. Space, Ocean	6(1)	1	0	0
10. Material	6(1)	2	1(1)	0
11. Labour	7(1)	4(1)	2	0

( ) : from the parent committee

(Source: [STA tf])

The most striking peculiarities of the STA Delphi study occur at Step 1 and Step 2B. At Step 1, the Committee is composed of as many as 16 members plus about 5 for each of the 11 subcommittees, the total number being 75 in the latest Delphi study. This number is more or less the same throughout the three studies. At Step 2B, the surprisingly great number of experts is mobilized as the respondents. It amounts to approximately 2,000 in the latest study (Table 3.3). This number is also nearly stable throughout the studies with a slight decrease (Table 3.4). Contrastingly, American representative articles on the Delphi method seem to assume implicitly that the number of respondents is much less [Dalkey 1968, Dalkey 1969] and some [e.g., Dajani et als] explicitly assume that it is ten or fifteen. This number is smaller than the number of the parent committee members in the STA Delphi study.

Table 3.3 DELPH Study Respondents, in 1982 (Final Round)

FIELD	OCCUPATION					AGES						SEX		JOB		Total
	Firm	Univ	Pub	Assoc	Journ	20s	30s	40s	50s	60s	over 70	M	F	R&D	Not R&D	
Energy, Resource	56	52	25	46	2	0	9	38	95	34	5	181	0	120	61	181
Agr. Forest, Fish.	18	19	38	27	1	0	46	26	52	20	1	103	0	77	26	103
House, Educ.	15	24	29	17	5	1	11	26	36	15	1	75	15	35	55	90
Envir,Safety	13	19	61	33	3	0	5	47	56	18	3	129	0	72	57	129
Health, Med.	13	34	36	10	0	0	0	20	54	17	2	88	5	63	30	93
Life sci.	37	29	23	5	0	0	6	34	46	8	0	91	3	83	11	94
Urban	69	44	38	18	6	0	16	71	70	17	1	174	1	95	80	175
Transport	28	20	36	38	2	2	13	35	49	19	6	123	1	60	64	124
Commun, Inf.	87	25	47	35	4	1	23	85	74	12	3	197	1	112	86	198
Space	35	25	30	27	0	1	9	32	58	15	2	117	0	71	46	117
Ocean	19	28	34	19	0	0	7	39	33	20	1	100	0	69	31	100
Materials	48	46	64	21	3	0	12	80	74	15	1	181	1	164	18	182
Prod. Labor	49	25	20	44	3	0	8	39	69	22	3	133	8	83	58	141
Total	487	390	481	340	29	5	123	572	766	232	29	1,692	35	1,104	623	1,727

firm: private firm, pub: government office and research inst.,
assoc: public or semi-public corporation, journ: journalist or writer

(source: [STA tf])

Table 3.4 Number of Respondents in Delphi Studies
(Final Round in Each Study)

Studies	Year	Sent out	Returned	%
1st	1971	3,108	2,482	80
2nd	1976	2,296	1,922	84
3rd	1982	1,962	1,727	88

(Source: [STA tf])

The reason for the great number of respondents is partly the historical situation. In the first STA Delphi study in 1970 - 1971, there were few specialists in the TF field. Hence PB felt the difficulty in selecting some nonexperts from many. In such a situation, selecting all of those who satisfy a certain criterion is a natural solution to this difficulty. The great-many respondents method exhausts almost all leading researchers, council members of the related professional societies, the directors of leading institutes and so forth on an almost mechanical criterion.

Immediately after the first STA Delphi study, it was pointed out [Eto 1972] that the STA Delphi study by the great-many respondents method is a kind of national technology assembly composed of the government-selected assembly-members and can be characterized as an aid to decision on technology policy rather than the forecasting. Though nobody supported this view, its actual development seems to have proceeded along the line of this view. As was stated just above, the great-many respondents method was a result from the lack of TF specialists in 1970. This method was not regarded as a new invention because it is a natural extension of the traditional collective decision by many participants commonly seen in many Japanese organizations. Now the situation has changed. Today there are a number of experienced TF specialists in Japan. But STA still continues the great-many respondents method. This fact leads to the recognition that STA considers the participation of many respondents as being more effective than the opposite.

Scientific judgement is usually believed more accurate if it is done by a fewer number of higher experts. The great-many respondents method necessarilly allows that some of the respondents are not highly trained for the judgement and tends to decrease the degree of accuracy of judgement. Political decision is usually believed more persuasive if it is done by a greater number of people who need not be experts. The idea of democracy excludes the superiority of decision by a fewer number of higher experts over that by greater number of people who need not be experts. The STA Delphi study with the many participants is more like the decision than the forecasting [Eto 1972].

The report of the latest STA Delphi study [STA tf] states that the convergence of experts opinions is attainable due to the tendency that respondents of relatively little specialty follow the majority opinion. This surprisingly honest confession indicates that the accuracy of forecasting is not necessarily important to STA. This is contrasting to the view stated in the representative articles by American Delphi specialists [Hills & Fowles] [Murray] which is concerned with reliability and accuracy of the Delphi forecasting. Reliability and accuracy are very important for the astronomical prediction or the weather forecasting but not so in policy making.

Accuracy is measured by comparing the forecasting f made at time t_0 with the result r at time t_r where $t_0 < t_r$. This measure is quite meaningful for predictions of natural events on which a sequence of decisions has no effect. Contrastingly to the natural events, the sequence of decisions d_1, d_2 ....., d_m made at t_1, t_2, ..., t_m respectively which determines the resource allocation ($t_0 < t_1 < t_2$, ... $t_m < t_r$) has a significant effect on the result r at t_r (Figure 3.1). The first decision d_1 is sensitively affected by the forecasting f meanwhile the later decisions d_2, ..., d_m are less dependent on f. If r is wholly dependent on f, the accuracy of f can be measured by the comparison of r with f. But the dependence of r on f is quite indirect via d_1, d_2, ..., d_m. Generally the concept of accuracy is irrelevant to the decision. When the forcasting is not independent of the decision but is an aid to decision, the concept of accuracy may not be highly relevant to the forecasting.

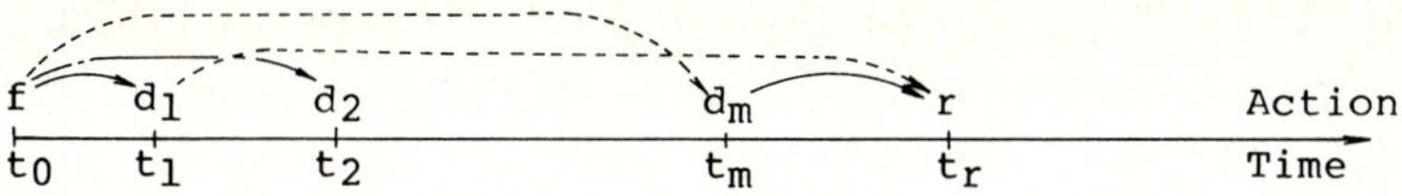

Impact, ——— : strong, —— – —— – —— : medium, - - - : weak

Figure 3.1 Impact of Decisions on Result

Analogy to accuracy may hold with the reliability. Scientific judgement is more reliable when it is made by a smaller number of higher experts whereas political decision is more acceptable when it is made by a greater number of people. When forecasting is considered more decision-oriented, more priority may be placed on societal or political acceptability rather than scientific reliability.

The reliability of the Delphi study is considered to depend on the stability of the responses between the rounds [Dajani et als.], [Hill & Fowles], [Nelson]. Indeed, the scientific statement will lose reliability if the statement varies. It is because science is concerned with the nature whose basic structure is stable in time. Different from the nature, society is considered quite changeable in time and a decision is demanded to adapt to social changes. Hence the stability itself is not a key factor for a reliable decision.

Another peculiarity of the STA Delphi study occurs at Step 3 where the number of rounds is predetermined. This termination criterion is very different from, or contradicting to, what is assumed as the termination criterion by many American Delphi specialists [e.g., Dajani et als.] who assume that the rounds proceed till the consensus is attained. This peculiarity results from the great-many respondents method which incurs a huge monetary cost for STA and time for experts who are busy persons. This situation requires STA to cut the number of rounds and, further, the rigid system of the governmental budget requires STA to predetermine the total cost in advance. Contrary to the idea of Dajani et als, the number of "rounds" in the political decision is often predetermined: the two-"rounds" system of legislative decision between the Lower and Upper

Houses and the three-"round" appealing system of judicial judgement. When the government (e.g., STA) sponsors the Delphi study, the sponsor may incline to the political manner of thinking on the round system. This may give rise to a significant difference from the Delphi study sponsored by a scientific or research organization (e.g., RAND Corporation).

The truncation of the iteration at the second round makes the instability of responses between the first and the second round indistinguishable from the convergence toward the consensus. So far the responses converge toward the consensus from the first to the second round for every case of the Japanese governmental Delphi studies, but nobody can tell thereafter. This effect of the truncation may be viewed as a demerit by the objective-truth seekers but as a merit by the political-acceptability seekers.

The selection criterion of the respondents is often criticized [Hill & Fowles], [Murray]. The biasness of the selection usually occurs when the sponsor of the Delphi study checks personality of the candidates for respondents. The great-many respondents method makes this kind of personal biasness impossible. In fact, they are selected on an almost mechanical criterion as was stated. As for biasness in the selection criterion itself, a degree of biasness of this kind is not unacceptable as far as the STA Delphi study is concerned because the selection criterion in the STA Delphi study is almost the same as that of the advisory boards of the government (except for the average age). The great-many respondents method makes it possible to extend qualifications to younger generation.

A third peculiarity of the STA Delphi study occurs in the questionnaire design at Step 2A. The questionnaire asks not only the year of breakthrough but the degree of importance (choice among minor, medium and major degrees), the national measures (financial aid, training of personnel, organizing or noncommitment), the way of promotion (self-reliance, technology import or international joint project) and the leading organs (public sector, industry or public-industry cooperation) for each issue of questions. These items are irrelevant to forecasting itself if the forecasting is more of the astronomical prediction type but is highly relevant to forecasting as an aid to policy making.

Issue	The breakthrough of X
Experties in the issue	High Medium Low
Importance of the issue	High Medium Low Unimp.
Time of breakthrough	- 85, 86 - 90, 91 - 95, 96 - 00, 01 - 05, 06 - 10, 10 -
Reason of after 10	Techn. Public accept. Econ. Others
Way of promotion	Self Import Int'l coop.
Major promoter	Public Joint Private
Measures by governm.	Fund Training Organ. Others No

Figure 3.2 Questionnaire Used in 1982 Delphi Study

(Made from [STA tf])

The superiority of group judgement over individual one is considered as an important condition for the usefulness of the Delphi method [Dalkey 1968], [Dalkey 1969], [Hill & Fowles], [Murray] since its origin. This superiority is measured as the convergence of responses toward the truth in a psychological experiment by Dalkey. Indeed, superiority of this kind may be one of the reasons why the collective decision system is often chosen in many organizations. But many social scientists are inclined to believe that more important reason for choosing the collective decision is its attainability of the equilibrium solution among conflicting subgroups. The concept of truth may be less important than the concept of equilibrium among conflicting interests to the policy making. Hence it seems that the governmental office prefers group judgement to individual one irrespective of its superiority with respect to the convergence toward the truth. From this point of view, the great-many respondents method is quite preferable.

It is likely that private firms implicitly feel the result of the STA Delphi study as indicating the STA policy rather than a forecasting itself though nobody explicitly express that. This feeling is quite natural for the above-stated reasons.

3.3 Indirect Coordination of Industrial R&D by STA

The TF Report by STA states that the STA Delphi study's purpose is to assist the formation of science and technology policy inside the government and to provide an orientation of R&D activities in the industry by publishing its result. The first part of this statement is related to the view in the subsection 3.2 and the second part is related to the view in the subsection 3.1 of this Chapter. As the part of assisting the science and technology policy formation was discussed in detail in the subsection 3.2, the second part is discussed again here with a special emphasis on the role of the STA Delphi study.

The STA Delphi study is sometimes the only TF result available to most firms. Therefore many firms are forced to rely on its result. More specifically, many firms choose one or two particular approach(es) among several alternatives according to the result of the Delphi study. Similarly, many firms set target dates of their projects according to it. In this way the STA Delphi study tends to unify the approaches of different firms and to synchronize the target dates of the technologically related projects under the independent management. This is a very strong coordinating power of STA.

The publication of the result of TF study has a strong effect but this is not the only effect. Another channel may be more important. Many scientists and engineers join the committees and become respondents from firms as Tables 3.2 and 3.3 show. Though only the limited number of firms send their staff as STA Delphi study members, they cover the technologically major firms and a significant portion of the whole industrial R&D resources. Thus, STA gains the direct channel to the industrial R&D.

Respondents from private firms are selected directly by the sponsor of the study (by STA in the case of the STA Delphi study) on the criterion of their scientific activities in related professional societies without prior recommendation of their firms. Accordingly they respond to the questionnaire as individual experts with no conscious reflection of the corporate interest. But their responses reflect, if not consciously, the actual state of R&D activities in their firms. When they revise their opinions in view of the fedback information, they tend consciously or unconsciously to revise their R&D activities in their firms in consistence to their revised opinions. In this way R&D activities in different firms are implicitly coordinated through the participation of individual personnel in the study beside the publication of the result.

Implicit coordination works analogously among R&D activities in public research institutes and universities. There is no single organ to coordinate all of these organizations. The Act for Establishment of Science & Technology Agency explicitly excludes the universities from the scope of the STA functions and implicitly excludes the industrial R&D by limiting the scope to coordination of the governmental administration perhaps for the laissez faire principle in the universities and the industry. Despite the absence of the unifying organ for formal coordination mechanism among these research organizations, the nation-wide STA Delphi study plays a role of coordination among the separated organizations.

Figure 3.3 illustrates in what way the nation-wide Delphi study reflects the actual state of R&D activities and affects their reviews of the on-going projects and their plans for new projects in different organizations. Figure 3.4 illustrates its effect on the national unity.

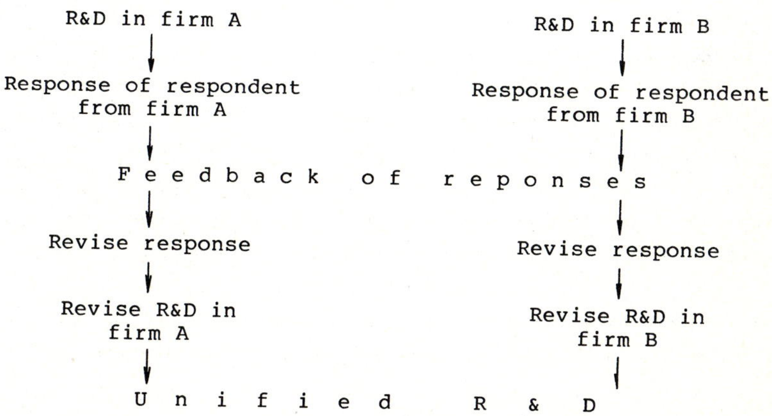

Figure 3.3 Behaviour of Delphi Respondents

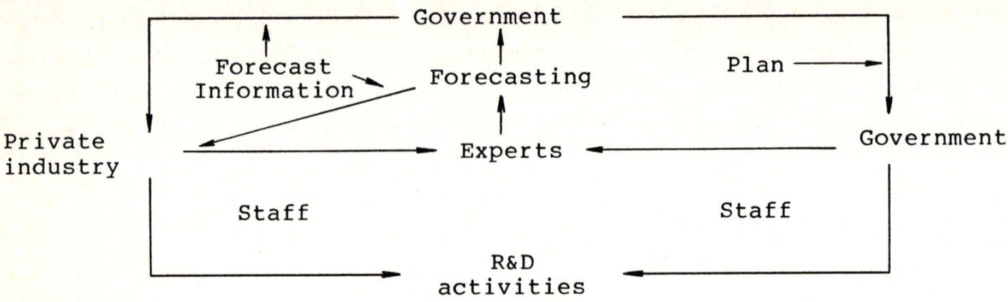

Figure 3.4 The Pattern of Government-Industry Cooperation

3.4 Direct Intervention of STA in Seed-Creation

STA has its own research institutes in basic and small-science fields like metals and inorganic materials beside big-science fields. Since materials science is considered as one of the most promising key fields, R&D activities in these institutes and the STA policy based on them are expected to stimulate the entire R&D activities.

STA undertook a remarkable program in 1981 for promoting the radically creative science and technology by growing the seeds for radical breakthroughs. Four projects started in 1981, each with 2 or 3 billion yen budget over the planned term of five years and each with 100 - 181 million yen in 1981. This program is entrusted to the STA-supervised special corporation JRDC (Japan Research & Development Corporation) which collects the project leaders from industry and universities (for the first four projects, one from industry, two from university-attached research institutes and one from a university). The project leaders then collect researchers (about twenty for each project) from industry, universities and public research institutes on a contract basis (for about five years). This flexible system is taken with no "brick and block" investment allowed.

The management system of this program is quite different from the prevailing Japanese system in the following points: (1) The project leader collects researchers on a contract basis in this program. Usually, on the other hand, the leader (professor) selects assistants in universities as careers instead of on a contract basis. In large private research institutes the selection is usually made by the personnel department with limited consideration of opinions of the project leaders. (2) The theme of research in this program is defined at first for clarifying the target, but it is so flexible that its unexpected development into a new theme is highly encouraged. Contrastingly, the target is unclear and development into an interdisciplinary theme is often discouraged in universities. Meanwhile the target is often too rigid in many private research institutes.

This new program by STA adds to the relatively new program by the Ministry of Education, each having the contrasting concept. The Ministry of Education (Figure 3.5) set up the National Inter-University Research Institutes (NIURI) for making radical breakthroughs in basic sciences (high-energy physics, basic biology, molecule science and physiological science) in this decade though the administration of R&D by the Ministry of Education was almost limited to the university and the university-attached research in stitutes till 1972. NIURI is based on traditional "brick and block" investment and in-house research by permanent career researchers plus university professors as opposed to the contract-based flexible activities of JRDC. The purpose of NIURI is defined as assisting the university research by the Act for the Establishment of the National Universities and Schools amended in 1973. In fact their cooperation with the industry [Inose et als] is limited to cowork for laboratory plant design instead of the collaborative R&D itself. NIURI may be compared to the research institutes under the Academy of Sciences in East European nations except for the administrative aspect, the former being under the direct administration of the Ministry of Education while the latter being autonomous and independent of the government. After the conflict with the Science Council, the government declined autonomous R&D organizations like the Academy of Sciences, and furthermore the Ministry of Education hesitates the collaboration between the universities and the industry as a reflection of the scientifically conservative university atmosphere.

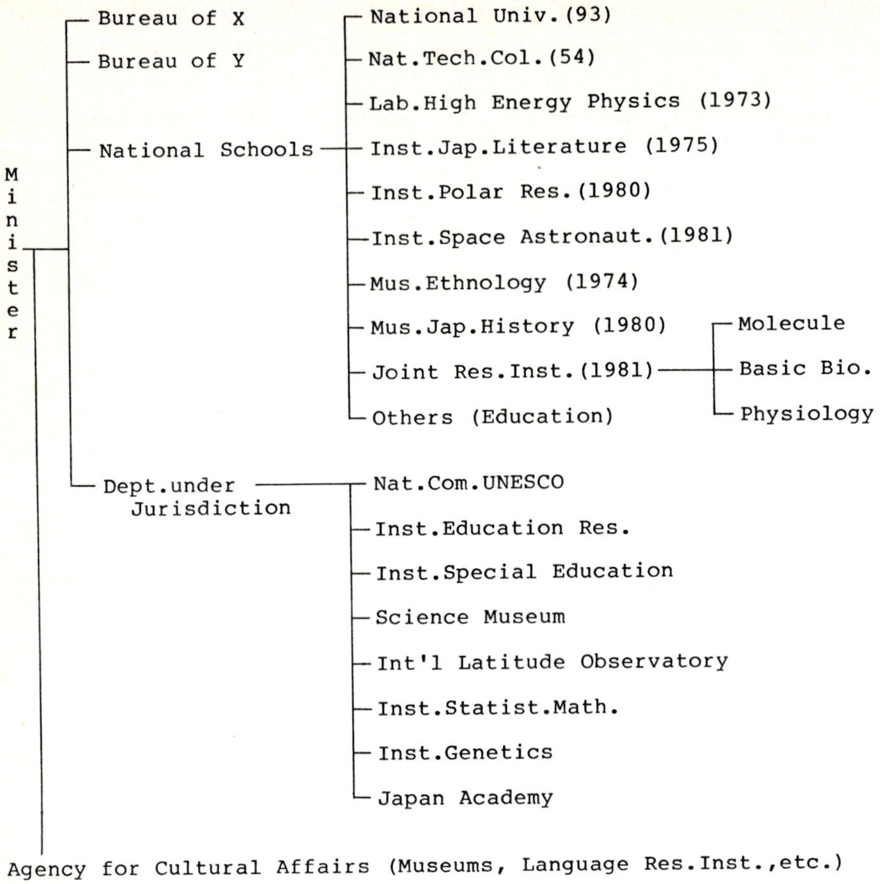

Figure 3.5 Organization Structure of the Ministry of Education (made from booklets of Ministry of Education, 1982)

4. INDUSTRIAL R&D COORDINATION FUNCTION OF MITI

4.1 Position of Industrial R&D in Whole R&D

There are the various means for the government to promote R&D activities in the industry. The most typical one is that the government orders and purchases high-technology weapons like missiles, aircrafts, radar systems and so on. In fact, this was the most important one in Japan till 1945. Present Japanese technology in shipbuilding, automobile and communication equipment owes to the tradition of military technology. Today, Japan lacks powerful armed forces for legal and domestic-political reasons. Hence this means can not work today. Another means is that the government establishes and supports the public research institutes. This one has not been chosen in Japan for a historical reason.

Historically the Japanese government has not been interested in R&D. Before World War I, the role of the universities was to absorb Western knowledge and to teach it to students. There was no public research institute at first. Instead, there were only the national inspection laboratories, experiment stations, testing laboratories and so on. After World War I, several basic research institutes were founded by private funds. After the great economic panic around 1930 financially choked these private research institutes, the army and the navy took the leadership role in R&D promotion. After World War II, the army and the navy were disorganized and no organization took their place in R&D promotion. The financial situation was extremely difficult at that time. The persistent disagreement between the Science Council and the government started then. This made the government quite reluctant to R&D funding. A doubt prevailed whether technology can be developed without military stimulation. When Japan achieved the economic reconstruction in the second half of the 1950's, technology was already at the stage of big sciences like nuclear power and space science. The cost for big sciences was too heavy for the just reconstructed economy and, further, another doubt prevailed whether big sciences can be developed without military motivation. The lack of consensus on defence gave rise to a lack of consensus on big sciences which had been tied with defence. The recognition that these big sciences lead the whole R&D in the major nations discouraged national promotion of R&D.

International trade competition forced the industry to promote R&D. With very little support from the government, the significant portion of the whole R&D resources has been accumulated within the industry. Table 4.1 shows that the industry shares a larger portion of the whole R&D resources in Japan than in the other nations. In such an R&D resources distribution structure, it may be wise for the government to restrict its promotional activities to mainly utilizing the industrial R&D resources within private firms.

Ministry of International Trade & Industry (MITI)

- Other Councils
- Bureaus
- Industrial Technology Council
- Agency of Industrial Science & Technology (3965 persons)
- Japanese Industrial Standards Committee

Agency of Industrial Science & Technology

(Headquarters)

- General Coordination Department (228 persons)
  - General Coordination Division — Comprehensive coordination of the Agency planning of industrial technology administration,Secretariat of Industrial Technology Council, etc.
    - Office for Promoting Research into Technology for Resource Survey Satellites
    - Computer Utilization Technology Office
    - Office for Technology Development of Resource Survey Satellites
    - Energy Technology Planning Office
  - Personnel Affairs Division — Personnel affairs, welfare
    - Office for consultation for personnel transfer to Tsukuba
  - Budget and Accounts Division — Accounting, Equpment, Administration of Assets
  - Research Administration Division — R&D programs for affiliated Laboratories and Institutes, Coordination with other public research institutes, patent management
    - Industrial Science and Technology Advice Office
  - Planning Division — Expansion of research laboratory facilities
    - Office for Tsukuba construction
    - Office for Research Information, Administration, & Planning
  - Technology Promotion Division — Support for R&D in private industries through subsidies, tax exemptions government loans. Research Association for Promotion of Mining and Manufacturing technology. R$D on medical nad welfarae equipment technology.
  - Technology Research and Information Division — Survey of technological affairs, technology assessment, pbulication, library.
    - AIST Library
  - International R&D Cooperation Div. — International cooperation in Industrial Science & Technology
  - Research Coordinator for Prevention of Industrial Pollution — Planning for R&D on pollution-preventing technology at affiriated laboratories and institutes. Contact with public and private testing and research organizations

(Research Institutes and Laboratories)

- National Research Laboratory of Metrology
- Mechanical Engineering Laboratory
- National Chemcial Laboratory for Industry
- Government Industrial Research Institute, Osaka
- Government Industrial Research Institute, Nagoya
- Fermentation Research Institute
- Research institute for Polymer and Textiles
- Geological Survey of Japan
- Electrotechnical Laboratory
- Industrial Products Research Institute
- National Research Institute for Pollution and Resources Office
- Government Industrial Development Laboratory, Hokkaido
- Government Industrial Research Institute, Kyushu
- Government Industrial Research Institute, Shikoku
- Government Industrial Research Institute, Tohoku
- Government Industrial Research Institute, Chugoku

Unit	Function
Officer for Future Industrial Technolgoy	R&D for future industrial foundation
Councillor for Technological Affairs (in charge of large-scale industrial technology development)	
Senior Officer for Development Program (General Coordination)	General coordination on large-scale industrial technology development.
Senior Officer for Development Program	Large-scale industrial technolgoy development (recycling of resources, complex manufacturing, optical measuring and control, sub-sea oil, C_1 chemistry, manganese nodule, high-speed computing system, automatic-sewing system)
Councillor for Technology Affairs	
Senior Officer for Devleopment Program (General Coordination)	General coordination on new energy technology development
Senior Officer for Devleopment Program	New energy tehnology development (solar, coal, geothermal, general)
Senior Officer for Development Program	Energy conservation technology development (MHD power generation, hgih efficiency gas turbine, new type of batteries)
Tsukuba Adminsitration Office	Management and administration of joint facilities at Tsukuba Research Center, staff welfare.
Standards Department (93 person)	
Standards Division International Standards Office	Comprehensive coordination of standards department, surveys, propagation of JIS and JIS Marks, Japan Industrial Standards Committee, International Standardization
Material Standards Division	Industrial standards for basic common items in metal, civil engineering, construction, mining, etc.
Textile and Chemical Standards Div.	Industrial Standards for chemicals, textiles, ceramics, daily necessities, pakaging
Machinery Standards Division Transportation and Machinery Standards Office	Industrial Standards for machinery, aircraft, ships, railway carriers, automobiles, industrial vehicles.
Electricity Standards Division Reliable Technology Development Office	Industrial Standards for electronics, electric telecommunication, medical equipment, combustion equipment, nuclear power, R&D on reliability of electronic parts.

(Source: [AIST a])

Fig. 4.1 Organization Structure of AIST

Table 4.1 Percentage of Private-Funded R&D in the whole R&D Expenditure

Year / Country	1965	1970	1975	1979
Japan	69.1	74.7	72.4	72.6
France	30.2	37.0	40.5	41.6
FRG	52.6	53.4	51.0	53.2
UK	47.7*	48.3	48.3	51.9**
USA	35.1	43.4	47.0	51.0

* 1966 ** 1978

(Source: [STA i, 1981], [STA w 1978])

4.2 Planning Power of MITI

Legally, no ministry or agency is authorized to lead industrial R&D within private firms nor are they allocated sufficient amount of budget to do so. Traditionally the Ministry of Finance has been skeptical of the effect of R&D and still now its allocation of the university budget is mainly for education. This behaviour is quite natural under the condition that the top (the Prime Minister) takes no initiative and leaves decision to each minister. Hence the task of organizing the industrial R&D resources within firms for national goals is forced to be done with a limited budget and with no legal authorization nor formal mechanism.

Among the several ministries and agencies involved in R&D, the Ministry of International Trade and Industry (MITI) has the closest contact with the industry. The historically accumulated know-how of MITI for guiding the industry enables MITI to perform this task in a better way than the others. But this is not the only asset of MITI for this task. Another asset may be its organizational structure.

MITI is the only ministry that has a unified administration office specializing in R&D management for its own research institutes. The other ministries and agencies have no single office to administer their research institutes. As in Figure 4.1, MITI has Agency of Industrial Science & Technology (AIST) which administers all of the eight research institutes and laboratories of MITI. The Ministry of Welfare and the Ministry of Agriculture, Forestry & Fisheries have a greater number of research institutes and laboratories. The former has ten and the latter has fourteen. But neither of them has a unified office to administer their research institutes and laboratories. These ministerial research institutes and laboratories are under the direct administration of each minister. But this means that nobody is responsible for R&D administration under the Japanese management system as was stated in the subsection 2.5. Unlike the university research, the governmental research institutes are not the laissez faire system, but the lack of unified administration prevents the systematic promotion of R&D in all ministries except MITI. The difference in the organizational structure enables MITI to have stronger coordination power than the others.

AIST has a strong head office. It has Coordination Department with 228 personnel (5%) out of the total 3,965 personnel of AIST including secretaries and other assisting personnel. The significant number out of the 228 personnel have research experiences in either of the research institutes or laboratories. Interestingly, Coordination Department has a personnel rotation system with the research institutes and laboratories. Several researchers of age around the 30s are 'drafted' from the research institutes or laboratories to the head office for a year or so and serve as staff. This resembles the staff system in the army and the navy where staff were also around the age 30s and were subject to the regular rotation between the staff office and the field or front, but this is quite different from the staff or advisory system generally seen in the Japanese offices (the subsection 2.5). The mixture of researchers in active service in the administration office activates planning tasks and makes plans acceptable to researchers in the ministerial research institutes as well as private ones.

Another merit of AIST is its widest-scope of the territories among the industrial R&D research institutes. It ranges from mechanical and chemical engineering to fermentation. Even Hitachi Ltd. which has the largest R&D resources among the Japanese firms is weak in chemical and biological technology as Figure 4.2 reveals. Other technological giant firms like Mitsubishi Heavy Industry, Toshiba (Tokyo-Shibaura) Electric and so on have the similar drawbacks as Figures 4.3 - 4.5 illustrate. This superiority of AIST over any single private firm contributes to its distinguished planning fuctions.

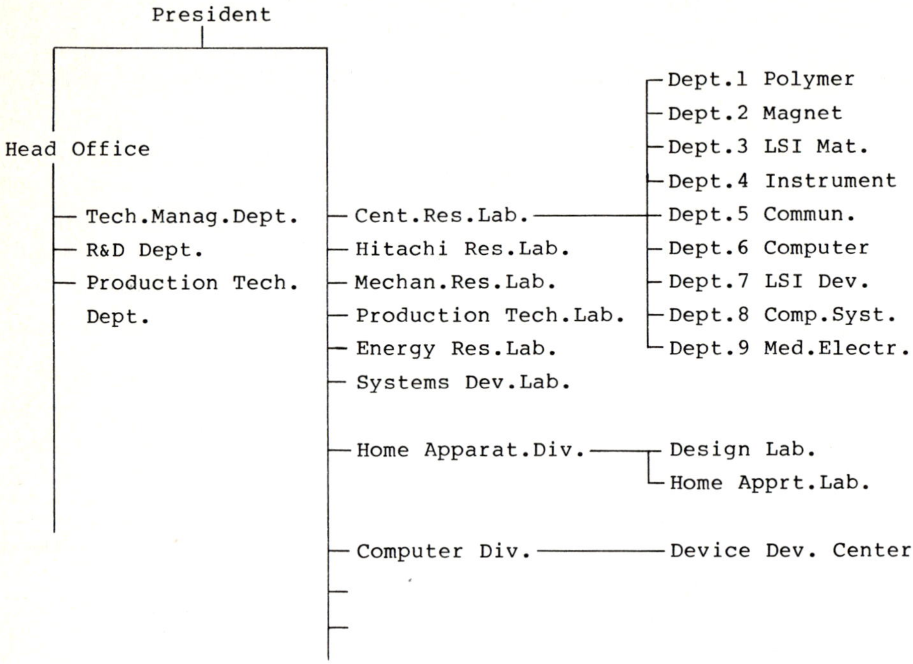

(Made from booklets of Hitachi)

Figure 4.2 R&D Organization of Hitachi

- President
  - Technology Head Office
    - Tech.Manag.Dept.
    - Development Dept.
    - Production Tech.Dept.
    - Patent Dept.
    - Nagasaki Res.Lab. (Shipbody, Engine)
    - Takasago Res.Lab. (Atomic Power Plant)
    - Hiroshima Res.Lab.(Naval Architecture)
    - Yokohama Res.Lab. (Disposal Plant)
    - Nagoya Res.Lab. (Heating & Cooling)

(madc from booklets of Mitsubishi Heavy Industry)

Figure 4.3 R&D Organizations of Mitsubishi Heavy Industry

- President
  - Head Office
    - Tech.Manag.Dept.
    - Production Tech. Dept.
  - General Res.Labs.
    - Metal & Ceram.Lab.
    - Chem.Mat.Lab.
    - Electron.Dev.Lab.
    - I.C. Lab.
    - Electron.Eq.Lab.
    - Inf.Syst.Lab.
    - Energy Plant Lab.
    - Mech.Lab.
  - Production Tech. Lab.
  - Division
    - Electronic Tech.Lab.
    - Medical Tech.Lab.
    - Semi-Cond.Tech.Lab.
    - Home Apprat.Tech.Lab.
    - Atomic power Tech.Lab.s
    - Heavy Electric Tech.Lab.

(made from booklets of Tokyo-Shibaura Electric)

Figure 4.4 R&D Organizations of Tokyo-Shibaura Electric

- President
  - R&D Headquarters
    - R&D Adminsiter. Dept.
    - Cent.Res.Lab.
    - Product Develop.Lab.
    - Consumer Electr.Develop.Lab.
    - Manufacturing Develop.Lab.
    - Consumer Products Res.Lab.
    - LSI R&D Lab.
    - Inform.Syst. & Electr.Develop.Lab.
    - Materials Eng.Lab.
    - Industrial Design Center

(made from Booklet of Mitsubishi Electric)

Figure 4.5 R&D Orgnizations of Mitsubishi Electric

4.3 Large-Scale Project System as Symbolic Grant

The most powerful leadership of MITI in the industrial R&D is performed in the form of large-scale projects. It may give the impression that MITI orders high-technology commodities from firms and purchases these directly from them as the Agency of Defense does. But MITI has no authority to purchase anything except for its own use. MITI is one of the poorest ministries in amount of budget allocated and has the least authority for purchasing. The Ministry of Construction is authorized to construct trunk roads including bridges and tunnels and so forth and is therefore allocated huge budget for these works. Hence the Ministry of Construction could promote R&D in construction companies through, for example, ordering modern bridges which require new technology. The Ministry of Education is allocated the university budget and purchases laboratory equipments through the universities by reallocating the budget to the universities. Hence the Ministry of Education could promote R&D, for example, in medical opto-electronics by establishing interdisciplinary departments in medical schools and allocating budgets for purchasing new equipments. On the other hand MITI is allocated no budget for purchasing large-scale commodities.

The themes of the MITI's large-scale projects are shown in Table 4.2 wherein none of the products can be purchased by MITI itself. Most of them may be purchased by firms and some may be purchased by other governmental organizations. The resource recovery system which is one of the themes of the large-scale projects may be purchased by a local government which administers the disposal system. But the local governments are not supervised by MITI but by the Ministry of Home Affairs. In fact none of the projects has been commercially successful in the sense that nobody purchased the products themselves except only one case. The only one exception is the sea water desalination project whose product has been exported with the international market share being as high as 3/7 in 1979 [STA w, 1982]. The internationally strong competitivity of Japan in this field highly owes to this project.

Table 4.2 Large-Scale Projects

(in million yen)

Project name	R&D period (FY)	Total R&D expenditure	Outline of project
Resource recover technology	1976-1982 (2nd phase 1973-1975 (1st phase)	about 11,300 1st phase: about 1,300	R&D on technical systems for the disposal of solid urban waste, centered on resource recycling with a view to promoting the efficient utilization of resources and facilitating the smooth application of solid urban waste treatment.
Flexible manufacturing system complex provided with laser	1977-1983	about 13,000	R&D on complex production system in which mechanical components for small batch production of diversified products can be flexibly and rapidly produced from metallic materials in an integrated system.
Optical measurement and control system	1979-1987	about 18,000	R&D on measurement control system that used optics and makes possible the systematic measurement, integrated observation and control of mass information, arising in specific areas, such as industrial parks, large-scale plants, etc., even under adverse conditions such as those where electromagnetic induction, etc., prevails.

(to be continued to the next page)

Project name	R&D period (FY)	Total R&D expenditure	Outline of project
C(mono-carbon) chemical technology	1980-	about 15,000	R&D on technology for the stable and economic production of such basic chemical products as ethylene glycol, acetic and ethanol, etc., with C_1 compounds such as carbon monoxide obtained from alternative carbon sources such as natural gas, coal, tarsand, etc., as their basic raw materials.
Manganese nodule mining system	1981-1989	about 20,000	R&D on efficient and reliable mining systems using hydraulic pump and/or air lift pump for deep-sea deposits of manganese nodules which contain economically important metals (Ni, Cu, Co, Mn, etc.)
Scientific computing system	1981-1989	about 23,000	R&D on high-speed computer system for processing and computation of scientific and technological information (processing of image information from satellites, simulation of nuclear fusion, etc.) which present computers cannot handle in realistic time.
Automated sewing system	1982-	-	R&D on automated sewing system that is composed of several processes such as preparation, making-up and finishing, for the apparel industry faced at diversification and rapid change of domenstic demand.
Subseal oil production system	1978-1984	about 15,000	R&D on subsea oil production system (in which production from small-scale oil fields is possible without any adverse effects on the fishing industry) which is effective for subsea oil production in deep waters (over 300m deep) and suitable for use in the sea areas surrounding Japan.
Super-high performance electronic computer	1966-1971	about 10,000	Large scale computer system with super-high performance.

(to be continued to the next page)

Project name	R&D period (FY)	Total R&D expenditure	Outline of project
Desulfurization process	1966-1971	about 2,600	(1) Efficient removal of the SO_2 contained in the grass exhausted from power plants or other industries which consume a great deal of heavy oil. (2) Direct removal of sulfur from heavy oil.
New method of producing olefin	1967-1972	about 1,100	Economic production of olefins by direct cracking of crude oil instead of using naphtha.
Remotely controlled undersea oil drilling rig	1970-1972	about 4,500	Remote-control oil drilling rigs for undersea use.
Sea water desalination and by-product recovery	1969-1977	about 6,700	Economical large-scale production of fresh water and economical by-product recover technology.
Electric car	1971-1977	about 5,700	Various types of electric car to replace ordinary vehicles in urban areas.
Comprehensive automobile control technology	1973-1979	about 7,300	Integrated control technology with a view to relieving traffic congestion, reducing automobile pollution and traffic accidents, etc.
Pattern information processing system	1971-1980	about 22,000	Computer technology for the recognition and processing of pattern information such as characters, pictures, objects and speech.
Direct steelmaking process using high temperature reducing gas	1973-1980	about 13,700	Direct steelmaking technology with a view to solving the pollution problems that accompany present-day methods and reducing the dependence on coal as a raw material. The new technology aims at a closed system which uses the heat energy from a multi-purpose high temperture gas-cooled reactor in the steelmaking process; this new reactor is scheduled for development in the near future.

(to be continued to the next page)

Project name	R&D period (FY)	Total R&D expenditure	Outline of project
Olefin production from heavy oil as raw material	1975-1981	about 13,800	Technology for manufacturing high-value-added olefin (commonly known as ethylene, propylene, etc.) using a high sulphur-content heavy oil fraction (so-called asphalt), which is difficult to desulphurize, as the raw material.
Jet engines for aircraft	1976-1981 (2nd phase)	about 13,000	Research and development of large scale turbofan engine designed for the use in commercial transports in the 1980s.
	1971-1975 (1st phase)	about 6,900	

(Source: [AIST a])

MITI claims that the purpose of the large-scale projects is the promotion of R&D as a technological side-effect. The purpose of the aero-space technology projects by NASA or the Department of Defense in USA is not limited to technological side-effects but rather the primal purpose is to produce and operate the systems developed. Accordingly, the purpose for American firms to join projects is not limited to technological side-effects but rather their primal purpose is to sell the products and related services. The major motivation of participating firms in USA is profit rather than the technological side-effects which is merely a side-effect to the firms anyway.

Japanese firms also desire commercial success. For this desire to come about, the project would desirably be sponsored by respective ministries (the disposal system by the Ministry of Home Affairs and the coumputer-controlled transportation system by the Ministry of Transport and so forth). But the situation that the other ministries lack the planning function forces firms contented with the commercially unsuccessful projects of MITI.

The procedure of the MITI's large-scale projects may be described as follows. First, MITI (actually AIST) plans the projects and decomposes a project to detailed subtopics. Then individual firms applies under appropriate subtopics. AIST reviews the applications and selects one firm for each subtopic. The selected firm is awarded the grant.

The limited amount of budget allocated to MITI naturally limits the amount of the grants. Generally, the grant covers only some part of the cost and the firms must pay the remainder of the cost from their own funds. The grant is merely symbolic and carries little financial meaning in an extreme case. Nevertheless patents obtained in the projects are usually owned by MITI for the reason that the research is done under a grant from MITI though the firm which contributed to obtaining the patent is given the privilege in utilizing the patent.

A question may arise as to why firms join the project even when the grant is so small. This question may be answered by considering the decision behaviour in firms which will be discussed in the Section 5.

4.4 Intercorporate Cooperation under Loyalty to Company

It is widely known that Japanese employees are highly loyal to their companies. It is also known that Japanese employees in different firms cooperate with each other in joint projects. These seemingly contradicting facts may be explained by governmental coordination across the firms. But this explanation leads to another question about the basic ground of the successful governmental coordination. The latter question may be answered, as far as the cooperation of engineers across the firms is concerned, by strong unity of alumni organizations in the professional fields.

Personal ties through the alumni were very strong in the Japanese society till the mass education destroyed them. Mass education is typically seen in the economics and business departments wherein hundreds or even thousands of students meet in a class. This prevents the development of alumnus ties among the graduates from these departments. The aim of these departments in the Japanese universities is to provide huge numbers of businessmen to the industry. As the graduates from these departments comprise the majority of businessmen in the industry, the discussion on the behaviour of the Japanese businessmen [e.g., Tsuda] is exclusively concerned with them. With no other community in particular, they form friendly circles inside the company and have strong identification with the company. But this discussion [Tsuda] fails to consider the behaviour of engineers.

The personal ties through the alumni is still strong in the medical community because the man-to-man education system is well preserved in the medical schools and the frequent occasion of reunion at the professional societies strengthens their unity. As the man-to-man education system is relatively preserved in the engineering colleges and as most of engineers join professional societies, the analogy holds with the engineers world. The graduates from the economics and business departments usually join no professional society and even refuse to be the specialists as described in the subsection 2.5.

Personnel departments in Japanese firms manage engineers differently from the economics and business departments graduates because their behaviours are very different. In almost every company the employees are explicitly classified into the clerks and the engineers and different personnel management systems are applied.

Engineers who studied in the same department of the same college or university have the strong unity in the same generation as well as across different generations. The engineering education system in the Japanese college and university provides the poor program for the general education at the freshman and sophomore levels and lacks the system of minor specialities beside the major at the senior and graduate levels. Resultingly, Japanese engineers at the time of joining a company have a very strong inclination to the major specialty trained in the colleges and have a very strong identification with alumni and professional societies till retraining programs and the rotation systems of the company wash their brain. The reputation that the Japanese engineers in the company have a wide-scoped view [e.g., Bradbury et als] may be attributed to the personnel management of the company rather than the college education.

The personnel management of companies, of course, does not deny the joining of the engineers to professional societies and, rather, often encourages it for updating their technological knowledge. On the occasion of meetings of societies they strengthen the professional unity as a side-effect. This unity helps their cooperation beyond the 'wall' of firms when they join intercorporate projects. It is positively noted that professional societies played a decisive role in many intercorporate technological cooperations in the shipbuilding, camera and steel industries which now own the world-top technologies [Eto 1979A].

Professional societies are not the only kind of organizations for promoting intercorporate technological cooperations. Perhaps more important one is the so-called semi-public & semi-private organizations. Some functions and authorities of MITI or other ministries are entrusted to these organizations wherein retired governmental officials and company employees, and rotated governmental officials and company employees work together. The funds of the semi-public & semi-private organizations are from both of the government or government-related bodies and industry. They receive contracts from both of the government or government-related bodies and the industry. For each of contracts a study group or a committee is organized with members from both of the government or government-related bodies and the industry.

For an example of such semi-public & semi-private organizations, let the Japan Industrial Technology Association (JITA) be considered here. JITA was set up in 1969. Its funds came from another MITI-related semi-public & semi-private organization and the industry (900 and 300 million yen respectively at its start). Its top executive was (also is) from the industry. The majority of its high executives and middle managers were (also are) from the industry with a few important management positions from MITI. Its major purposes were (also are) to coordinate between public and private research activities for avoiding overlap and to make government-owned patents readily available to private firms. As one of the tasks derivated from these purposes, JITA receives contracts from AIST of MITI. It has very few number of in-house researchers but it collects researchers from AIST and the industry for each contracted research topic and forms an ad hoc committees or working group for the topic.

Some of the subtopics of the MITI's projects are contracted out to and done by these organizations. These parts are usually called the adhesive parts in the divided topic. Also these organizations themselves are often called the adhesives or the tirtiary sector between the public and the private sectors.

Engineers join committees or working groups with the consent of the company. They can often receive some secretarial assistance or permission of using a computer or other facilities of the company for this project. The company earns valuable technological information at these invisible and negligible cost. Usually, companies are relatively willing to send their engineers to such contract research outside the companies.

The theme of the large-scale projects is divided into small subtopics each of which is done separately by a distinct firm in most cases. The semi-public & semi-private organizations coordinate between firms besides the direct coordination by AIST of MITI. This coordination works on the ground of the professional unity of engineers across firms. Figure 4.6 summarizes the discussions above.

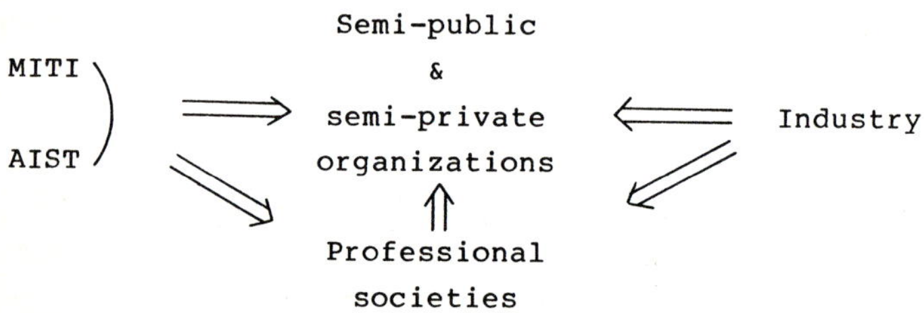

Figure 4.6 Intermediation between MITI and Industry

These semi-public & semi-private organizations naturally incur some cost. In fact, one of the aims of these organizations is to create jobs for retired governmental officers. Knowing this, the press critisizes the part of the national budget for them especially at times of economic recession. The long-lasting recession since 1974 forced the Cabinet to promise disorganizing some of them and to set up a special task force with a prominent chairperson for this purpose. It is still in the hot debate but the Cabinets were forced to declare it as one of their most urgent and critical policies in the 1980's. If these organizations are really disorganized, the worldly wellknown cooperation between the government and the industry will be subject to radical changes.

5. GOVERNMENTAL INTERVENTION IN INDUSTRIAL R&D DECISION

5.1 Responsibility Aversion in Bottom-up Decision System

The Japanese decision system is often called the bottom-up decision system. More precisely, it is the middle-up decision system in that collective decision by middle managers is almost straightly authorized by the top. The major functions of the top are to raise the morale of the middle managers and to monitor the decision process so that it works in a fair manner.

The collective decision is almost synonymous to the agreement of the majority. The collective decision system excludes the strong initiative of a project leader or an R&D management specialist and leads to participation of all the concerned managers. The majority of middle managers lacks the wide-scoped perspectives and tends to make a decision from a local or short-ranged concern due to their limited ability and experiences. This drawback is multiplied by their limited responsibility and limited allowance of risks. This leads to "decision by the elimination method" in that alternatives with no authorized evidence of certainty are eliminated and the rest with authorized evidence is chosen. In other words, alternatives with no particular guarantee of certainty are eliminated from the candidates list and the one which has survived this elimination process wins the support of the middle managers.

The decision by the elimination method is not exactly the same as the decision by the risk aversion principle. An authorized endorsement assures no success. In the case of the MITI's project, the planning ability of AIST of MITI is not perfect enough to assure even the technological success. If the American or European project is successful in producing a system with a better performance by using completely different approach, the experience with the MITI's project may be useless to the firm and it incurs the opportunity cost to the firm which has lost a better opportunity by joining the MITI's project. Furthermore, MITI is in no position to guarantee the commercial success of the project as was described in the subsection 4.3. In fact, all of the MITI's projects with an exception are commercially unsuccessful as a natural result of the legally limited power of MITI. Knowing these facts, the middle managers are still inclined to the alternative with the MITI's support. This behaviour may be called responsibility aversion in that they decline to take responsibility of choosing another alternative in competition with the MITI-endorsed alternative. By choosing the MITI-endorsed alternative they leave the responsibility to MITI and thereby they are excused from the responsibility if the project chosen fails.

The responsibility aversion is the risk aversion under the demerit system in personnel management. Personnel management in many firms takes the demerit system in evaluating their personnel. The demerit system is a personnel management version of the elimination method in that personnel with a demerit or a failure is eliminated from the candidates list for the promotion to higher position and personnel with no or few demerit survives this elimination process. Hence the middle managers tend not to take the responsibility of decisions for avoiding the risk of demerit.

Responsibility aversion leads to selection criterion of projects on which the R&D theme, as a part of the MITI's project, is given priority despite no real guarantee of technological and commercial success. The grant from MITI covers only some portion (often less than 50%) of the cost but this does not hinder this selection. The problem is not to avoid the financial risk of the company but to avoid the promotion risk of the middle managers. The grant is merely a symbol in that it covers only some part of costs but it symbolizes that the middle managers can be free from a possible demerit if the project chosen results in failure. Figure 5.1 illustrates the typical decision behaviour in the Japanese firms.

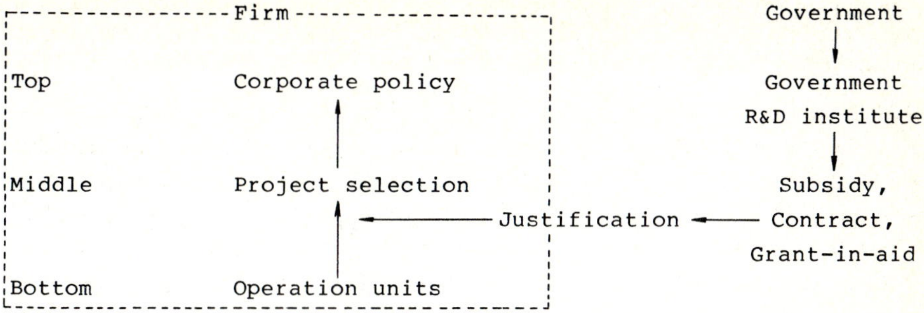

Figure 5.1 Justification in Process of Project Selection

5.2 Role of Initiator of MITI's Projects

The typical investment behaviour may be expressed in terms of leader and followers in association with speculation. When stagnation starts, nobody wants to take a risk of initiating a new project till somebody undertakes the venture. The role of public investment is often to initiate investment with the expectation to warm the cooled investment mind.

Many scientists tend to choose the waiting strategy till somebody courageously advances into a new research front. This risk averse behaviour of scientists is amplified by the responsibility averse behaviour of their managers. To break the waiting strategy, somebody must stir the stagnation. MITI plays the role of initiator while firms are just waiting for the stimulus.

The large-scale projects of MITI may be of little technological importance but its role of initiating R&D investment may be more important. Today the intervention of MITI into the industrial R&D is severely criticized by the other industrialized nations. If MITI accepts this international criticism and withdraws from the intervention, the R&D decision system of firms may have to be radically changed.

5.3 Gate Keeper in Long-Range Planning

The Japanese top manager (the Emperor, minister, company president and so forth) devotes most of his time to ceremonial tasks. He acts as a morale monitor in the course of decisions as was stated in subsection 5.1. He is supposed to keep quiet till his men reach a consensus. But this does not mean that he need not attend the meetings. His attendance raises the morale of the participants who judge the degree of importance of the meeting by his attendance. Also his visit to the factories and research laboratories symbolizes that the activities in these factories and laboratories are important to the company. The engineers measure the degree of importance of their tasks in terms of the degree of his attention to their tasks. Similarly his visit to governmental offices and customers are extremely important. Usually his visit is merely a good-will visit after the material business is already done by his men. But this does not allow him to escape from this time-consuming task. This prevents him from having enough time to think of the long-range strategy. Middle managers usually lack long-range perspectives, and Japanese firms and governmental offices lack the excellent staff organization. A question arises as to who thinks of the long-range strategy.

First of all, the government plays the essential role of forming the long-range technology development plan for firms in terms of the MITI's project or so. But another question is as to how the MITI's plan is acceptable to firms. Traditionally, MITI has a strong channel with the industry and MITI exploits this in guiding the firms. But an important question of intermediator between MITI and the firms has been ignored in the discussion on the MITI-industry link [e.g., Tsuda] presumably because the discussion so far has been exclusively concerned with the routine management of financial tactics rather than the long-range technology strategy. This may be extremely important for the long-range technology strategy formation.

Various kinds of people work in a company. The majority of Japanese businessmen are exclusively concerned with the internal matters of the company. These people are called companymen by the press. But there are some exceptional people. These exceptional people are more curious about what is going on outside the company and want to see what is beyond routine jobs. These people are quite informed of discussions in the governmental offices, professional societies and the press.

The governmental policy is based on public opinion which is led by the experts, journalists, columnists, writers and so on. The opinion leaders take the leadership role in forming public opinion which in turn influences the governmental policy formation process through the advisory boards or the press. The opinion leaders also directly affects the governmental policy by joining advisory boards themselves (Figure 5.2). The direct influence of journalists on the government is seen in the structure of advisory boards. Even the highly specialized committee of technology forecasting of STA includes a popular science fiction writer, a popular modulatress in TV programs, a technology newspaper writer and a medical columnist among the 16 committee members. In this way, the opinion of influential journalists is a leading indicator of future governmental policy or governmental guidance to the firms on long-range strategy even in the technology field. Needless to say about the influence of experts on the governmental policy and guidance. The gate keeper of a firm has a sensitive antenna for the opinions of these leaders through their publications by journalism, lectures at meetings of professional societies and proposals at governmental advisory boards. The gate keeper translates these opinions into the jargon of the firm and prepares the framework by which the governmental guidance is to be accepted. In other words, the gate keeper plays the role of the "fifth column" of governmental guidance in advance of guidance itself.

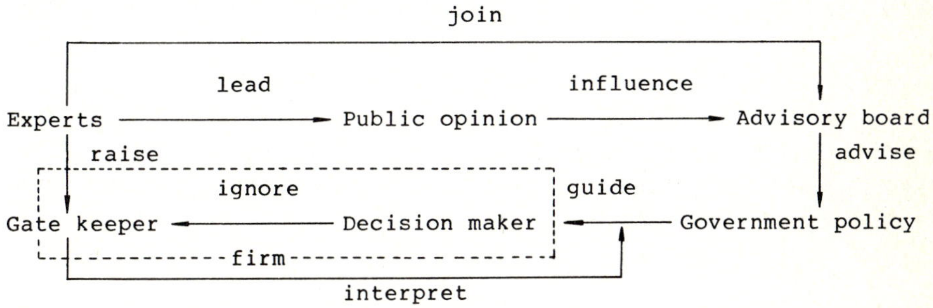

Figure 5.2 Reflection of Public Opinion to Firms Strategy

The gate keeper is not intentionally trained by corporate personnel management. Management even wants to remove him as a non-cooperative person to the present corporate policy in a similar way that the fifth column is oppressed in the war. But the life-time employment system prevents his dismissal and the gate keeper or the fifth column spontaneously grows and prepares for governmental intervention into the corporate planning.

Despite persistent oppression, the gate keepers survived in the Japanese government offices and even the army and navy during the World War II and kept the window, if not the gate, open to the Western culture. When the Allied Powers took control of the Japanese governorship in the postwar period, the General Headquarters (GHQ) of the Allied Powers found the very useful assistants who understood the democratization policy of GHQ almost perfectly. The "fifth column" can survive even in wartime.

Recently some firms began to recognize the role of the fifth column as a keeper of the firm's gate to the public. Some firms now have employees with special talents who often work on intercorporate projects and in semi-public & semi-private organizations or join governmental committees. Though the merit of such gate keepers is today still not widely recognized nor highly evaluated by personnel departments, his contribution to forming the long-range strategy in harmony with national policies is not to be neglected.

6. INDUSTRIAL BEHAVIOUR TOWARD FUTURE INNOVATION

6.1 International Criticism over Japanese Science and Technology

The discussion in the foregoing sections was concerned with the domestic conflict and the national mechanism for resolving it. The discussion in this section will turn to the recent international conflict between Japan and the Western nations and the possible mechanism for resolving it.

The recent international criticism over the rapid excess of export of Japanese industrial products gave way to an international suspicion that Japan, in particular the Japanese government, places unfairly high priority on export-oriented technology at the sacrifice of a balanced system of science and technology. Or a more extreme version of this suspicion is that the Japanese export offensive is the other side of her poor innovativeness and that Japan enjoys a merit of followers. This suspicion may readily develop into another acute international conflict. This section will examine the following hypothesis and discuss the result of the examination.

Hypothesis 1.1 Japan is a follower rather than an innovator.

This hypothesis may be broken down into the following sub-hypotheses.

Hypothesis 1.2 Japan is not innovative.
Hypothesis 1.3 Japan makes little effort in basic research (BR).
Hypothesis 1.4 The Japanese industry is not active in BR.

6.2 Innovativeness of Japan

Hypothesis 1.2 in the subsection 6.1 may be broken down further into Hypotheses 2.1 - 2.4.

Hypothesis 2.1 The ratio of technology export-import of Japan is inverse to that of the Western nations.

Discussion 2.1 Table 6.1 shows that the ratio of Japan is inverse to that of France, UK and USA but is similar to or even better than that of FRG. The grouping between the export-excess nations and the import-excess nations is quite similar to the grouping between the "colony"-holding nations and the "colony"-not-holding nations. This interesting coincidence may be explained by a fact that firms often export their technology merely to their subsidiary firms abroad in a "colony" for local productions. It may deserve noticing that Japan now has a significant amount of excess of technology export (nearly double of import) for newly contracted cases and that her excess of import in the total is merely due to the continued cases. This trend may merely reflect her recent increase in local productions abroad for avoiding the overexport of products. Hypothesis 2.1 is anyway rejected.

Table 6.1 Technology Trade in Selected Countries, 1977 - 1980
(in billion yen)

	France			FRG			Japan								UK			USA		
	Rec A	Pay B	A/B	A	B	A/B	New A_1	Cont A_2	Sum A	New B_1	Cont B_2	Sum B	A/B	A_1/B_1	A	B	A/B	A	B	A/B
77	477	353	1.44	90	219	0.41	36	57	93	17	173	190	0.49	2.15	170	139	1.22	1269	120	10.6
78	-	-	-	91	203	0.45	47	75	122	38	154	192	0.64	1.23	157	132	1.19	1235	128	9.6
79	-	-	-	108	233	0.46	52	81	133	27	214	241	0.55	1.94	190	147	1.29	1357	155	8.8
80	-	-	-	-	-	-	74	85	159	28	212	240	0.67	2.68	-	-	-	1557	174	8.9

(made from [STA i, 1982], [AIST t])

<u>Hypothesis</u> 2.2 Japan is inferior to the Western nations in patenting.

Discussion 2.2 The direct balance in mutual patenting (balance between the patenting of Japan in a foreign country and the patenting of that country in Japan) is favourable for Japan against France, UK and USA. Japan is equal to FRG on this criterion. The patenting in a third country (patenting of country A in country C and that of country B in C) may be a better index to compare the performance of A and B [Soetes]. Table 6.2 shows that, for the number of applications, Japan surpasses many countries in many patent markets; Japan surpasses every major country in the patent market of USA, is surpassed only by USA in FRG and UK and is surpassed by USA and FRG in France. Table 6.3 for patent granted shows the exactly same result. The patent granted with the normalization by R&D expenditure (the Soetes' normalization) gives the similar result; Japan surpasses every major country in the patent markets of USA and FRG and surpasses USA in France and UK. Hypothesis 2.2 is rejected.

Table 6.2 Patents Applied by Nationality of Inventor (1980)

Applied country / Nationality of inventor	Japan	USA	FRG	UK	France	Canada	Switzer-land	Nether-lands	External Applica-tion (A)	Domestic Applica-tion (B)	(A)/(B) (%)
Japan	165 730	12 951	5 267	4 223	2 498	2 018	552	785	35 254	165 730	21
USA	10 391	62 098	6 606	6 953	4 644	13 125	1 139	1 688	73 756	62 098	119
FRG	5 761	9 669	28 683	3 560	4 082	2 148	1 837	1 086	46 189	28 683	161
UK	1 571	4 178	1 013	19 612	818	1 194	176	241	15 884	19 612	81
France	1 674	3 331	1 252	1 085	11 000	1 203	407	387	17 616	11 000	160
Switzerland	1 392	1 975	1 481	1 068	872	630	4 049	269	13 870	4 049	343
Netherlands	944	1 025	618	781	577	512	154	1 826	7 885	1 826	432
Canada	271	1 969	172	346	119	1 785	24	40	3 736	1 785	209
Application by foreigner (A)	25 290	42 231	19 900	22 028	16 989	23 189	5 613	5 532	-	-	-
All applications (B)	191 020	104 329	48 583	41 640	27 989	24 947	9 662	7 358	-	-	-
A/B (%)	13	40	41	53	61	93	58	75	-	-	-

(Source: [STA i, 1982])

Table 6.3 Patent Applications and Patents Granted by Country of Origin, in USA, FRG, Japan, France and UK in 1979

Country of origin	USA		FRG		Japan		France		UK	
	Applic.	Granted	Applic.	Granted	Applic.	Granted	Applic.	Granted	Applic.	Granted
USA	60535	30605	7226	3713	9864	4625	5806	5235	7777	5951
FRG	9261	4398	30879	10895	5277	1491	5035	4652	4530	3427
Japan	11185	5069	5296	2441	150623	34863	2656	1551	4049	2101
France	3128	1533	1780	1114	1536	509	11303	6846	1528	1278
UK	4270	1841	1451	715	1522	526	1164	1386	19468	4182
Canada	2061	892	203	114	238	101	164	132	397	221
Switzerland	1836	956	1874	958	1296	451	1193	1237	1198	844
Italy	1480	561	1178	345	642	157	1241	719	1052	388
Sweden	1137	561	813	347	604	194	505	445	695	405
Netherlands	983	500	948	581	888	407	850	764	965	546
Australia	579	203	136	41	227	34	96	70	295	153
USSR	494	329	492	353	390	242	470	333	242	250
Austria	447	214	556	211	174	80	227	162	187	128
Belgium	373	173	165	99	135	56	220	250	195	126
Israel	235	80	61	7	29	6	36	25	52	21
Finland	231	76	201	35	108	17	111	53	164	68
Denmark	223	108	195	101	105	52	131	69	233	107
Hungary	175	59	183	67	76	34	102	61	117	55
Spain	163	49	138	29	65	20	170	121	117	34
Norway	152	82	76	46	49	23	40	35	81	44
GDR	134	2	555	100	107	8	155	100	196	60
New Zealand	94	18	19	2	26	4	16	6	69	17
Poland	83	26	102	40	30	18	58	30	59	36
Mexico	82	35	7	9	15	6	7	14	16	6
Czechoslovakia	80	51	139	42	48	42	65	83	80	56
Brazil	72	19	31	2	19	2	821	6	20	4
Argentina	62	76	10	2	10	1	8	8	15	5

(Source: [Soetes & Wyatt])

Hypothesis 2.3 The managerial innovation is more important than the scientific and technological one in Japan.

Discussion 2.3 This is of course a matter of subjective judgement. The majority of R&D managers in major firms judge that the process innovation is the most important (Table 6.4). Hypothesis 2.3 is rejected.

Table 6.4 Contribution of Innovations to Business in the Late 1970s

Category \ Degree	much (%)	medium (%)	no (%)	total (%)
Process innov.	32	48	20	100
Product innov.	24	55	22	100
Managerial innov.	11	56	33	100

(Source: [STA w, 1982])

A general belief that the process innovation is inferior to the product innovation gives rise to the following hypothesis.

Htpothesis 2.4 Most of Japanese innovations are minor.
Discussion 2.4 Table 6.5 shows that the number of past important innovations by Japanese inventors is not high. But it is not especially low if one considers her past poor surplus of high quality manpower which may be determined and hence represented by the workers productivity. R&D is possible only when a certain amount of manpower is allocated to R&D away from the direct production. This is possible only when the labour productivity is high enough to support and supply the R&D manpower of indirect production. The last columns in Table 6.5 show that the percentage of minor improvements in Japanese important innovations is lower than that in USA and that the row of Japan is quite similar to that of FRG. Hypothesis 2.4 is only weakly supported with some reservation. The concept of learning curve may be valid for the innovation in that the number and quality of innovations depend on the total (i.e., integrated over time) amount of industrial activities experienced in a culture. This may explain the excellent performance of UK

Table 6.5 Number and Index of Major Innovations in Postwar Period as Compared to Surplus Measured by Productivity

	Innov.listed by OECD	Innov.listed by Gellman	Productivity=GDP/Workforce				Radicalness (Gellman)%			
			1950	1960	1970	Average	Minor	Major	Breakthrough	Sum
USA	74(100)	319(100)	100	100	100	100	41.4	31.2	27.4	100
UK	18(24)	85(27)	50	50	53	51	4.4	40.0	55.6	100
FRG	14(19)	33(10)	36	51	67	51	36.4	50.0	13.6	100
Japan	4(5)	34(11)	16	25	49	30	38.5	53.8	7.7	100
France	2(3)	21(7)	43	55	71	56	11.8	64.7	23.5	100
Canada	0(0)	8(3)	81	86	89	85	50.0	0.0	50.0	100
Total	112(-)	500(-)	-	-	-	-	34.7	36.7	28.6	100

(made from [Inhaber & Alvo], [Rosenberg])

despite her small population in that UK has the longest experience of industry and can share the experience of the Europe Continent and the American one. The West Europe forms a cultural world and the North America forms another while Japan alone forms a small, culturally isolated world. This has been an unfavourable condition to Japan.

In conclusion Hypothesis 1.2 is not strongly supported and several evidences exist to disprove it.

6.3 Basic Research Effort of Japan

Hypothesis 1.3 on Japan's supposed disinterest in BR may be broken down into Hypotheses 3.1 - 3.5 to follow.

Hypothesis 3.1 The BR expenditure in Japan is low.
Discussion 3.1 The amount of BR expenditure in Japan is nearly equal to that in FRG and is higher than that in France and UK. Its ratios in the whole R&D expenditure and in GNP of Japan are not so low as those in UK and USA.(Table 6.6) Hypothesis 3.1 does not hold.

Table 6.6 BR Expenditures and its Ratios in Whole R&D Expenditure and GNP

Countries \ BR		BR (billion yen)	BR/GNP	BR/R&D
France	1977	382	0.37%	21.1%
FRG	1977	622	0.45	22.3
Japan	1978	593	0.36	16.6
UK	1975	227	0.33	16.1
USA	1977	1464	0.28	12.7

(made from [STA i, 1982], [STA w, 1982])

Hypothesis 3.2 The input and output of basic science in Japan is low.
Discussion 3.2 The recent ratio of basic science against engineering in terms of doctorate degrees granted is quite low, but this is

a result of the recent priority on the latter. Table 6.7 demonstrates that Japan has drastically placed priority on the latter in the 1960's though no longer so. As a result of the past priority on the former before the 1950's (see the discussion in the subsection 2.2), the number of publications in basic science by Japanese authors is quite numerous as seen in Table 6.8 where the nationality of a paper is identified by the affiliation of its first author. Contrary to the traditional priority on basic science, the modern trend is to place higher priority on engineering. Relatively young countries like USA and Japan have adopted this trend. Table 6.9 shows lower priority of USA on basic science. More precisely, the number of journals in basic science published in USA is not large despite the language advantage to attract contributions of English-writing scientists over the world. In addition, Table 6.10 shows a decline of the American share in the number of influential journals in basic science. Japanese contribution to physics measured in terms of the normalized citations by the Price's method is rather high as being greater than 1.0 for USA(1.01), FRG(1.05), France(1.23) and USSR(1.76) though less than 1.0 for UK(0.82) and Canada(0.81) [Burke et al] as Table 6.11 shows. The figure 1.76 as the entry at the place of USSR-Japan is the highest in Table 6.11. Hence it may be difficult to support Hypothesis 3.2.

Table 6.7 Doctorate Degrees Granted with Growth Rate for Japan (in person and %)

		Science	Engineering	Agricul-ture	Med,Pharm, Health
Japan	1965	416	419	241	2677
"	1970	611(46.9)	853(103.6)	353(46.5)	2662(-0.6)
"	1975	676(10.6)	986(15.6)	385(9.1)	2371(-10.3)
"	1981	822(21.6)	1186 (20.3)	527(36.9)	3537(49.2)
FRG	1979	2467	982	469	4827
UK	1978	2484	956	181	493
USA	1978	7374	2742	950	718

(made from [STA i, 1982], [ME])

Table 6.8 Papers in Chemical Abstracts 1979

Countries	Percent
USA	27.5
USSR	19.0
Japan	9.7
FRG	6.0
UK	5.9
France	4.2
India	3.1
Canada	2.9
Italy	2.2
Others	19.5
Total	100.0

(Source: [STA w, 1982])

Table 6.9 Percent of Journals in each Field from each Country

	Clinical Med.	Biomed. Res.	Biology	Chem-istry	Physics	Earth and Space Sci.	Psy-chology	Mathe-matics	Eng.and Tech.	All fields
US	21.0	22.2	19.0	25.1	19.7	17.2	40.4	15.5	19.9	20.1
UK	7.4	14.1	12.4	17.9	14.9	7.7	13.5	9.0	16.8	12.6
East & West Germany	8.8	7.8	5.7	8.7	9.2	7.2	10.3	9.0	8.8	8.0
France	6.6	5.4	4.3	2.8	4.5	5.3	6.1	4.8	5.1	5.2
USSR	5.0	12.3	6.6	9.8	16.5	12.1	0.9	8.7	9.8	8.8
Japan	7.8	6.8	6.0	5.5	5.4	4.4	2.9	4.2	6.4	6.1
Australia	1.0	1.4	3.3	1.2	1.4	2.7	0.7	0.8	2.3	2.1
Canada	1.5	1.2	3.4	0.8	0.9	4.7	2.0	1.5	2.0	2.4
India	1.9	2.0	2.3	1.2	0.7	1.2	2.5	2.2	1.9	1.9
Israel	0.1	0.0	0.3	0.4	0.0	0.3	0.0	0.6	0.2	0.2
Italy	6.6	3.4	2.8	2.6	3.8	2.6	1.6	3.0	2.3	3.4
Sweden	1.4	0.5	1.8	0.7	0.9	1.7	1.8	2.3	2.1	1.7
New Zealand	0.3	0.1	1.1	0.1	0.1	0.9	0.0	0.5	0.5	0.6
South Africa	0.4	0.1	1.2	0.4	0.4	1.2	0.9	1.1	0.7	0.8
Rest of East Europe	6.0	5.6	7.1	8.6	7.6	7.3	5.2	10.7	7.1	7.0
Rest of West Europe	14.0	11.2	12.1	10.1	11.0	13.4	8.5	16.2	10.7	12.1
Rest of Asia	2.0	1.5	2.5	0.8	0.6	2.6	0.4	2.3	1.3	1.8
Rest of Africa	0.9	0.3	2.0	0.2	0.3	2.0	0.2	2.1	0.3	1.0
Rest of Near East & North Africa	0.7	0.2	0.5	0.4	0.0	0.7	0.0	1.1	0.1	0.4
Central and South America	6.5	3.9	5.5	2.5	1.9	4.7	2.0	4.3	1.8	4.0
Total (approx.)	100	100	100	100	100	100	100	100	100	100

(Source: [Carpenter & Narin])

Tale 6.10 US Share of the World Publications from a large sample of Influential Journals, 1973-75

Field	1973 %	1974 %	1975 %
All fields	39	39	38
Clinical medicine	43	42	43
Biomedical research	39	38	39
Biology	46	46	44
Chemistry	23	22	22
Physics	33	33	32
Earth and space sciences	47	47	45
Engineering and technology	42	42	41
Psychology	76	75	75
Mathematics	48	46	44

(Source: [price])

Table 6.11 Normalized Citation Indicator in Physics (1973 and 74)

		Cited Countries							
		US	UK	FRG	France	USSR	Japan	Canada	Others
Citing countries	US	1.02	0.99	0.90	0.91	0.96	1.01	1.14	0.94
	UK	1.00	0.92	0.94	0.97	0.73	0.82	1.07	1.14
	FRG	0.96	0.99	1.05	1.20	1.06	1.05	0.77	1.12
	France	0.96	0.97	1.08	1.08	1.00	1.23	0.87	1.06
	USSR	0.98	0.86	1.29	1.19	0.96	1.76	0.69	0.95
	Japan	1.11	0.83	1.04	1.02	1.08	0.99	0.82	0.82
	Canada	1.11	1.12	0.72	0.87	0.66	0.81	0.74	0.89
	Others	0.93	1.09	1.23	1.12	1.26	0.96	0.96	1.09

(Source: [Burke & Price])

<u>Hypothesis</u> 3.3 Japan places low priority on the university research.

Discussion 3.3 The share of the universities in the whole R&D expenditure is the highest in Japan as shown in Table 6.12 which also shows the surprisingly many faculties in the Japanese universities much above USA. The Japanese input to the universities is enormous as far as the amount of expenditure and the number of faculties are concerned. Hypothesis 3.3 is thus rejected. The only remaining problem is their quality and performance.

Table 6.12 R&D Resources in Universities and Colleges

		Expenditure		Faculties
		Amount (billion yen)	% in the whole R&D	(thousands persons)
France	1977	284	15.5	22
FRG	1978	529	16.6	19 (1977)
Japan	1979	779	19.1	97
UK	1975	118	8.4	-
USA	1979	1570	13.2	78

(made from [STA i,1980,1981,1982], [AIST t])

Hypothesis 3.4 The number of Japanese Nobel prizers is low.

Discussion 3.4 The left columns of Table 6.13 support this hypothesis on one hand and indicates a language barrier against the non-English or non-Scandinavian countries like USSR, France and Japan on the other hand. The language barrier is low in mathematics. In fact the share of these non-English or non-Scandinavian countries for the Fields Medal, the most important prize in mathematics, is higher than that in the Nobel prize. The existence of the language barrier in the scientific area is demonstrated by many evidences. For examples, citations from the English-speaking countries is concentrated to the English-speaking countries [Burke & Price]; Japanese geographers tend to refer to English papers first and to the German papers second and so on [Haigh] exactly in the same order of their study of foreign languages; and only the very distinguished chemists measured in terms of their citedness are given the opportunity of international activities measured in terms of the editorship of international journals for Japanese, Russian and French [Zsindely et als]. These evidences casts a doubt that the number of Nobel prizers reflects the actual quality of basic science of a country. But Hypothesis 3.4 itself is valid.

Table 6.13 Number of Winners of Major Prizes

Prizes	Nobel (science only)		Fields
Year	1901 - 44	1945 - 82	1936 - 78
USA	19	109	8
UK	22	40	3
Germany (E&W)	35	14	0
France	16	6	4
Sweden	6	9	1
USSR	2	8	2
Switzerland	4	5	0
Japan	0	4	2
Belgium	2	3	1
Austria	6	2	0
Denmark	5	2	0
Italy	3	2	1
Netherlands	8	1	0
Others	7	15	2

(Source: [STA i, 1982] [STA w, 1982])

Hypothesis 3.5 Japan's funding in frontier fields is low where frontier fields denotes big science or sciences in space, nuclear energy and ocean.

Discussion 3.5 Table 6.14 demonstrates that the Japanese governmental funding in these fields used to be quite low but is high today.

Hypothesis 3.5 holds no longer as far as the recent funding is concerned. Interestingly, the oil crisis in 1973 and the 200-nautical miles territorial sea problem around 1977 gave quick rise to the fundings in nuclear energy and ocean development respectively. This fact indicates that the rapid increase of Japanese fundings in these fields was a response to the real needs in view of her deepest dependence on oil import and sea food among the industrialized countries. This may further allow one to generalize that the Japanese R&D funding behaviour reflects the industrial needs.

Hypothesis 1.3 may be difficult to support.

Table 6.14 Non-Military R&D Expenditure for Big Sciences (in billion yen)

	1973	1974	1975	1976	1977	1978	1979	1980	1981
Nuclear Energy									
Government									
France	158	187	213	188	167	150	191	226	-
FRG	101	121	142	143	147	143	184	200	-
UK	39	57	64	58	55	53	65	88	-
USA	415	505	391	429	734	556	607	677	-
Japan	63	73	104	120	144	175	198	248	-
Japan Non-gov.*									
Industry	15	19	21	26	44	35	36	40	-
Res.Inst.	61	64	67	82	102	143	151	176	-
Univ.	3	7	7	20	17	22	21	20	-
Total	79	90	95	128	163	200	208	236	-
Space									
Government									
France	56	56	67	70	69	64	85	102	-
FRG	61	66	62	65	62	62	77	87	-
UK	15	22	31	28	28	23	28	-	-
USA	872	838	897	990	959	795	922	110	114
Japan	37	58	77	88	95	96	100	102	105
Japan Non-gov.*									
Industry	5	4	8	3	5	6	11	8	-
Res.Inst.	23	39	59	69	56	69	74	74	-
Univ.	4	5	6	7	8	9	11	11	-
Total	32	48	73	79	69	84	96	93	-
Ocean									
Government									
France	11	13	11	11	10	10	11	15	12
FRG**	14	16	20	-	-	10	18	19	-
UK	5	7	9	6	10	-	-	-	-
USA**	172	194	221	259	253	224	246	242	248
Japan	11	11	14	17	20	32	33	45	60
Japan Non-gov.*									
Industry	5	5	5	6	5	7	9	9	-
Res.Inst.	6	8	9	10	11	12	15	16	-
Univ.	1	3	2	2	3	3	3	4	-
Total	12	16	16	18	19	22	27	29	-

* possibly include some governmental issue,
** include military budget

(Source: [STA i, 1978,1979,1980,1981,1982])

6.4 Seeds of Industrial R&D

Classically, R&D was done or at least led by the universities. Since the 1940's the governments have taken the lead in the whole R&D in the major countries. In this postwar system, the major R&D

themes are planned by the governments though some of them are actually done in the private sector on the contract basis. Exceptionally, the industry has been the leader of the whole R&D in Japan. A question may arise as to how the industry finds seeds or selects the R&D themes. Hypothesis 1.4 that the industry leaves BR to the others may be broken down into Hypotheses 4.1 - 4.2 to follow.

Hypothesis 4.1 The seeds of industrial R&D come mainly from universities, public research institutes and abroad.
Discussion 4.1 Table 6.15 shows that the overwhelming majority of the successful 63 innovations in 1975-78 are identified to have come from other domestic firms as the main source of their seeds. This fact rejects Hypothesis 4.1 and indicates that major innovations owe to inter-sectorial technology transfer or joint development. The rows of industry in Table 6.14 show that the industry does not follow the increase in the governmental funding for space and ocean development (note that nuclear energy is already inindustrial use and is different from these two) presumably because the industry does not really expect seeds from these fields. As to relatively small science like natural energy R&D, the R&D investment behaviour is quite similar between the industry and the universities (Tale 6.16). This fact makes it quite doubtful that the industrial R&D follows the University R&D with a timelag.

Table 6.15 Source of Seeds (single answer)

Sources	Answer
Other Japanese firms in different sectors	44.8%
Other Japanese firms in the same sector	34.5
Foreign firms	17.2
Japanese universities and colleges	3.4
Total (approximation)	100

(Source: [STA w, 1977])

Table 6.16 Researchers and Expenditures on Energy R&D by Energy Source

Energy sources	Total		Industry		Research institutes		Universities and colleges	
	Expenditures (million yen)	Researchers (persons)	Expenditures (million yen)	Researchers (persons)	Expenditures (millon yen)	Researchers (persons)	Expenditures (million yen)	Researchers (persons)
Fossil energy (total)	81 069	2 878	74 160	2 363	5 576	263	1 333	252
Oil	27 551	987	26 713	877	435	28	403	82
Gas	6 333	366	6 105	328	133	19	96	19
Coal	45 859	1 424	40 739	1 095	4 359	191	760	138
Other	1 326	101	604	63	648	25	74	13
Natural energy (total)	35 551	2 566	24 541	1 279	6 816	543	4 195	744
Geothermal	5 668	346	3 564	190	1 707	78	398	78
Solar	20 237	1 330	16 085	801	2 071	184	2 082	345
Marine	1 472	121	944	58	282	14	246	49
Wind	2 007	140	1 387	54	345	29	275	57
Biomas	5 580	567	2 419	168	2 134	207	1 027	192
Other	587	62	142	8	278	31	167	23
Nuclear energy (total)	235 826	6 589	39 582	1 983	175 776	2 532	20 468	2 074
Power plant technology	85 296	1 747	19 601	1 035	65 112	604	584	108
Multipurpose utilization of nuclear	7 047	181	1 303	77	5 583	78	161	26
Nuclear fuel cycle	39 325	644	7 156	271	31 655	300	514	73
Fusion	54 349	882	2 816	161	39 599	241	11 934	480
Other nuclear	18 486	828	5 990	247	10 960	437	1 535	144
Nuclear ships	4 689	111	134	17	4 535	89	21	5
Utilization of radiation	18 204	1 487	1 809	133	12 122	473	4 272	881
Radiation protection	8 429	709	773	42	6 210	310	1 446	357
Energy conservation (total)	134 571	8 376	121 879	7 043	8 209	653	4 483	680
Field of Industrial	41 879	2 501	38 885	2 053	1 619	221	1 375	227
Field of Public Welfare	31 185	2 357	30 554	2 253	328	44	302	60
Field of Transport	41 197	2 286	38 544	2 037	1 924	151	729	98
Conversion and storage of electric power	15 042	782	11 133	554	3 009	106	900	122
Hydrogen	2 989	284	1 452	74	834	85	702	125
Other	2 279	166	1 310	72	495	46	474	48
Other energy	14 004	736	5 107	325	8 094	297	802	114
Total	501 021	21 145	265 269	12 993	204 471	4 288	31 281	3 864

(Source: [STA i, 1982])

<u>Hypothesis</u> 4.2 The industry invests for BR as a follower of the public BR.

Discussion 4.2 This may hold true for Western countries, in particular for Protestant countries as is shown in Table 6.17 wherein the ratio of private BR expenditure in the whole private R&D expenditure is exactly concordant to that of public BR expenditure in the whole public R&D expenditure in ranking among the four Protestant countries. The Japanese industry makes a contrasting behaviour which seems to try to compensate for the insufficiency of public BR. In fact, Table 6.18 shows that the Japanese industry is increasing the expenditure for BR very quickly (as high as 27.6% in 1980). Hypothesis 4.2 is rejected as far as the expenditure or investment behaviour is concerned.

Table 6.17 BR in Public R&D Expenditure and BR in Private R&D Expenditure

		public BR		private BR	
		%	rank	%	rank
France	1977	8.5	5	3.2	4 (3)
FRG	1975	13.7	1	3.3	2 (1)*
Italy	1977	10.8	3	2.1	6 (5)
Japan	1979	5.5	7	4.6	1 (-)
Sweden	1977	5.7	6	1.8	7 (6)*
UK	1975	11.1	2	3.3	3 (2)*
USA	1977	9.4	4	3.0	5 (4)*

Correlation coefficient = -0.086 (includ. Japan), 0.435 (exclud. Japan)
(rank) : rank excluding Japan
* : Protestant countries

(made from [STA i, 1981])

Table 6.18 R&D Expenditure and BR Expenditure and Growth Rates

(in billion yen and %)

	Whole R&D (growth)			BR (growth)		
	Firm	Res.Inst.	Univ.	Firm	Res.Inst.	Univ.
1976	1882 (11.7)	470 (11.4)	543 (11.7)	95 (9.2)	78 (10.7)	306 (6.9)
1977	2110 (12.1)	487 (3.6)	576 (6.1)	99 (4.2)	85 (9.0)	330 (7.8)
1978	2291 (10.9)	561 (15.2)	656 (13.9)	106 (7.1)	99 (16.5)	376 (13.9)
1979	2665 (16.3)	600 (7.0)	686 (4.6)	123 (16.0)	110 (11.1)	378 (0.5)
1980	3142 (17.9)	682 (13.7)	714 (4.1)	157 (27.6)	104 (-5.5)	399 (5.6)

(made from [STA i, 1978,1979,1980,1981,1982])

Discussion 4.1 denies that the industry is a follower in BR in quality and Discussion 4.2 denies it in quantity. Hypothesis 1.4 is hence rejected. The Japanese industry which is the leader of the whole R&D in Japan is highly interested in BR.

The discussions above indicate that the industry relies or is trying to rely on its own effort in finding or generating the seeds. Then a question arises as to in and to which fields the industry is trying to find and apply the seeds. Discussion 4.1 already denied the possibility that space and ocean development is the major ground on which the seeds of the Japanese industrial innovation grow in the near future. The possibility of natural energy source (or soft energy path) as their major ground is also deniable as Table 6.16 shows that the amount of industrial R&D expenditure for natural energy is not big enough. The subsection 6.5 to follow will be devoted to identifying the major target fields of the coming industrial R&D.

6.5 R&D Priority in Industry

Science & Technology Agency (STA) publishes a series of books on the status quo in several important fields (see the subsection 3.1). The fields covered by this series reflect the interest of both the government and the industry. The fields covered are materials technology, science information systems (data base), ocean technology, life science and systems technology.

Ocean technology can be omitted from this discussion as was stated in the foregoing subsection. Science information systems may also be excluded because they are the task of JICST (Japan Information Center of Science & Technology) which is an STA-related public organization (see the subsection 3.1).

Systems technology as defined by STA in this series denotes an interdisciplinary policy analysis for social issues like the urban, environmental and energy problems. STA has named it "soft science" as denoting a system of knowhow to integrate the knowledges and technology of natural and social sciences for problem solving. Soft science or systems technology may be a small science as compared to space science in terms of budget needed but may be bigger than the latter in that the former includes social, behavioral and medical sciences as its essential components. This kind of bigness is a serious dilemma in the Japanese personnel management system which works well in a homogeneous personnel composition but works poorly in a heterogeneous one. The Japanese personnel management system separates engineers from the clericals (see the subsection 4.4) but soft science or systems technology defined by STA requires their mixing. As personnel departments dominate other departments (e.g., the accounting department) in most of Japanese firms today, firms tend not to establish soft science teams which personnel departments tend to avert. The tradition of the Japanese universities which hates interdisciplinary or mixed teams prevents the training of soft scientists or systemists and supporting the industrial BR in this field. Without support of universities most of firms can not afford to expand soft science or systems technology teams beyond a limited stage except giant firms. The lifetime employment system of Japan prevents firms from bringing in experienced leaders, if any, of interdisciplinary teams from outside who are unavailable in the firms.

As soft science or systems technology is closely connected to information technology and information systems, firms in the sector of electrical equipment and electronics are the most earnest in it. In fact, the central and local governments had joint research teams in this field with computer firms around 1970 in response to social needs at that time. But the economic recession in the mid 1970's cooled this market off because the recession caused the severe cuts of the public investment of large scale systems. In such a situation R&D managers of electrical and electronic firms noticeably shifted their priority from need-pulled innovations onto the seed-pushed ones (Table 6.19). This may be interpreted that systems technology or soft science is not the major target field of R&D in electrical and electronic firms which play the central role in this field.

Table 6.19 Priorities of R&D of Firms in the Past (- 1978)/future(1978 -)

(multiple answer, %)

Sectors (samples) / Priorities	All industry (626)	Synthetic fibre (59)	Steel (29)	Electric & Electronic (68)	Auto (27)
Need-pulled new products in the same field	71.6/ 53.8	64.4/ 52.5	69.0/ 62.1	88.2/ 54.4	85.2/ 66.7
Quality improvement in the same field	39.5/ 37.1	45.8/ 44.1	44.8/ 48.3	45.6/ 36.8	25.9/ 11.1
Energy saving in the same field	10.9/ 9.7	11.9/ 8.5	6.9/ 6.9	8.8/ 7.4	14.8/ 33.3
Resource saving in the same field	12.6/ 7.3	25.4/ 11.9	17.2/ 6.9	2.9/ 0.0	11.1/ 11.1
Labour saving in the same field	21.1/ 17.6	5.1/ 5.1	13.8/ 13.8	22.1/ 22.1	25.9/ 7.4
Seed-pushed new products in the same field	18.1/ 34.5	22.0/ 32.2	6.9/ 10.3	14.7/ 51.5	18.5/ 48.1
Shift to more basic fields	1.3/ 2.7	0.0/ 1.7	3.4/ 0.0	4.4/ 2.9	0.0/ 0.0
Shift to more applied fields	4.6/ 9.7	6.8/ 20.3	13.8/ 13.8	1.5/ 8.8	0.0/ 11.1
More R&D in different fields	2.4/ 5.1	1.7/ 6.8	10.3/ 0.0	1.5/ 2.9	3.7/ 0.0
No answer or Don't know	0.6/ 4.3	1.7/ 3.4	0.0/ 3.4	0.0/ 5.9	0.0/ 0.0

(Source: [STA p, 1978])

Table 6.20 shows that R&D managers in the electrical and electronic industry expect to achieve improvement innovations in the early 1980's rather than radical innovations. This may indicate that its major target field is not life science because the life-scientific innovations are radical to the electrical and electronic industry. In fact, Table 6.21 demonstrates that neither the electrical nor the electronic industry plans to reinforce R&D in life-science-related activities. (More precisely, none of the major firms plans to do so). This endorses the above-stated fact that only a few of electrical or electronic firms expect to achieve radical innovations. On the other hand, some firms in the electrical industry (note that about a half of the leading electronics firms are still classified in the electrical equipment sector for a historical reason) plan to reinforce R&D in basic chemistry. These evidences may be interpreted that the major target field of the electrical and electronic industry is materials technology rather than life science or bio-technology.

Table 6.20 Expected Innovations (single answer, %)

Sectors (samples) / Innovations	All industry (626)	Synthetic fibre (59)	Steel (29)	Electric & Electronic (68)	Auto (27)
Radical innov.in early 1980s	6.2	6.8	3.4	7.4	3.7
Radical innov.in late 1980s	7.3	11.9	0.0	8.8	3.7
Improvement innov in early 1980s	70.9	67.8	72.4	82.4	77.8
Improvement innov in late 1980s	12.3	13.6	24.1	1.5	11.1
No significant innov.before 1990	2.1	0.0	0.0	0.0	3.7
Not answer or Do not know	1.1	1.7	0.0	0.0	0.0

(Source: [STA p, 1978])

Table 6.21 Plan to Employ more Researchers for Selected Disciplines

Sector / Discipl.	Const	Food	S.fibre	Pharm	Rubber	Steel	N.fer	Machn	Elctr	Elcrn	Auto	Prec	Existing researcher in firms
Biology	△	○	○	○	×	×	×	△	×	×	×	×	2571
Pharmacy	×	◎	△	◎	×	×	×	×	×	×	×	×	5463
Basic chem	△	△	○	△	○	△	×	△	△	×	△	×	44487
Appl. phys	△	△	△	×	×	△	△	△	○	△	△	○	-

× : 0, △ : 1 ∿ 19, ○: 20 ∿ 39, ◎: 40 ∿ 100%

(made from STA p, 1981 , [STA i, 1982])

Table 6.21 also demonstrates that only a few of the industrial sectors plan the reinforcement of biological R&D. Set aside the pharmaceutical and food industries which have been deeply involved in biological R&D, the synthetic fibre, machinery and construction sectors are interested in biological R&D. The latter two are probably interested in bio-mechanics and human-factor engineering rather than biology in the narrow sense. Thus, the only remaining industrial sector which is deeply interested in the promotion of biological R&D is the synthetic fibre sector which now enscopes the various subfields of chemistry and is hence sometimes called the synthetic or integral chemical industry in Japanese word. The major territory of the synthetic fibre industry is considered to be materials technology rather than bio-technology or life science itself though they are in a close relation.

R&D in life science or bio-technology is under unfavourable conditions imposed by personnel management. As was stated above on soft science or systems technology, personnel departments in most of Japanese firms tend to avert the mixing of employees from different disciplines and hence tend to avert to employ biologists among the existing physicists and engineers. On the other hand, materials R&D can be done in the existing frame based on physical or technological knowledge systems and is not hampered by the aforementioned unfavourable condition stemming from personnel management.

Many firms also declined new and expensive investments in R&D especially after the oil crisis. Table 6.22 indicates that the cost concern was the major reason for the decline in R&D investment in the mid 1970's. Bio-technological R&D requires new facilities while the materials R&D can basically be done by the existing facilities. This is another favourable condition for materials R&D over bio-technological R&D.

Table 6.22 Reasons for Decline in R&D in Mid 1970's
(single answer, %)

Reason	%
Increasing risk due to increased R&D cost	28.0
Decreasing seed due to stalement of BR	24.0
Decreasing stimulus due to global stalemate of innovation	17.0
Decreasing incentive due to decreased inventor's profit causd by speed-up of technology advance	16.1
Diversification of need due to variety of social value	8.6
Decreasing acceptability of new technology to public	5.0
Miscellaneous	1.3
Total	100

(Source: [STA P, 1977])

The above discussions reveal that materials R&D is the most likely to be the target field of the Japanese industrial R&D among the above listed candidates. This prospect does not conflict with the recent trend of the increasing investment of the government for big sciences. Every issue of the White Paper of Science & Technology by STA stresses that the expected breakthroughs in big sciences can be achieved with the great aid of progress in materials technology in magnetics, alloys, membranes, new ceramics, new plastics, compound materials and so on. In this sense, advances in materials technology are expected to contribute to both of big sciences and industrial innovations.

6.6 Ideas underlying the Japanese Technology System

As a matter of fact, many Western people still feel that Japan merely follows Western technology. The discussions in the subsections 6.2-6.5 claimed that this impression does not reflect the reality. But this impression may be reasonable in the sense that the Japanese technology system is quite different from the Western one and is perhaps weak in some aspects where the Western one is strong. Of course this is merely a matter of criterion. As nobody can tell whether the Western poem is better than the Eastern one or not, the impression that the Western technology system is better than the others is logically groundless. Nevertheless the prevalence of this impression in the Western world may be grounded on the fact that no characterization has been properly done about the Japanese technology system (for an attempt, see [Eto 1979B], [Eto 1980A] , [Eto 1980B], [Eto 1982]).

The concept of technology has traditionally been opposed to that of spirit ("Geist" in German or the thought) in Japan for a century and been associated with Western culture. The balance between Japanese spirit and Western technology was ideal in the late 19c. Hence the Japanese felt no need to clarify the idea underlying the Japanese technology system or even has never thought of the existence of idea itself behind or within technology. But it will be one of the possibly significant contributions of Japan to the mankind to develop a new idea of a technology system for the 21st century type. Since her rejoining the international community a century ago, Japan seems to have made almost no attempt to justify her acts upon universally acceptable principles. Nearly a half century ago, Japan merely sent her army to East Asia without stating the idea of liberating Asia from the Western Colonialism or even without claiming to open the Asian colonial areas for the Japanese trade. Such an only military invasion only from her selfish desire for resource and market naturally resulted in strong anti-Japan sentiment in Asia. Today Japan merely sends commodities to the international market without explaining its expected effect on social innovations through accepting a non-Western technology system. The Japanese Ambassador to USA rebuted the US criticism of the Japanese closed industrial system by remarking that USA tends to judge any non-American way as being unfair (April 15, 1983). His remark itself may be true in that it reflects the reality but he seems to have forgotten that the US criticism is based on the universally accepted principle of free-market or the openness of international trade. Instead of proposing another universal principle to supersede the existing one, he merely proclaimed the national interest of Japan. It has always been the behaviour of Japanese diplomatic policy either to yield to Western Colonialism or to expound her own national interests without an internationalistic view of global cooperation for the mankind. In fact, the Japanese diplomatic history is a cycle which alternates between one-sided surrender to the Western Powers (before 1930 and 1945 - till recently) and one-sided claim of her national interest (1931 -

45 and today). Such lack of universal principles might have a negative effect on her resolution of international conflicts. The recent high-technology dispute between the Western Powers and Japan is no exception. At first Japan admitted, or at least made no counter-argument nor counter-measure to, the Western criticism that Japan is merely a follower. When the criticism began to threat Japanese exports, the Japanese attitude suddenly changed to maintaining her necessity to do it for self-protection.

It is true that Japan is merely a follower of the Western Powers both in invading East Asia and in employing unfair trade behaviour with less competitive nations. In fact, Western science started with the learning from Arabic one; East Asia was first occupied by the Western Militarism; and whaling was conducted to the near extinction of the species by the Western culture first. Japanese diplomacy and press have always tried to use these facts as excuses for her acts without attempting to replace the currently prevailing and accepted ideas with a new one.

For the purpose of resolving the recent internaitonal conflict over the Japanese trade supported by her innovation, the formal mechanism of diplomacy may not necessarily work effectively. As STA has been doing in domestic matters (see the subsection 3.1), an informal mechanism to deliver informations directly to the public may be more useful. In Japan discussions have been made on Japanese science and technology in the fields of traditional science & technology philosophy, modern pragmatic R&D management, R&D policy as a part of industrial policy, and journalistic reviews have been written for the domestic readers, but very few atempts have been made to conceptualize the modern Japanese technology system. For examples out of a few, [Shirane] and [Hayashi] are to be mentioned, but these were addressed to the Japanese public written in Japanese and need to be updated and internationalized.

6.7 Indigenous Basic Research and Technology

In a view of the history of science and technology, technology was puulled by industrial needs in preparation for and as a promotion of the industrial revolution in 18-19c. At that time many inventions were made by only moderately educated people. Only after economic wealth was accumulated enough, the wealth enabled some people to find time for scientific activities as an aritstocratic hobby. This kind of research pushed technology in 19-20c. Since the mid 20c. when the center of science and technology moved from Europe to USA, the government began to pull technology as the purchaser of high-technological weapons. Moreover, the government began to pull even the basic sciences as the sponsor of the basic research for atomic, biological and chemical (ABC) weapons and their carriers in space science including astronomy, nuclear physics and biology incuding medical science. Such scientifc activities for the destructive purpose have been justified as pioneering the bright frontiers for their expected technological side-effect on the industry, agriculture and medical clinics. This expectation tends to provoke the critical opinion that the nation, like Japan, which seeks exclusively for the industrial innovation is a selfish free-rider in scientific contributions to the mankind. But the deepening specialization widens the distance between disciplines or fields and makes military technology quite a separate one from industrial, agricultural and clinical technologies. This makes the transfer of knowledge from the former to the latters quite difficult [Eto 1979B, Eto 1980B]. The side-effect of the former on the latters is quite doubtful today [Eto 1980C].

In the age of high specialization and narrowed channels between disciplines or fields, knowledge acquired by BR within the industry (agriculture or clinics) may work more effectively for innovations there than knowledge acquired elsewhere. Such an "indigenous" BR is possible only when the military or governmental big science projects do not monopolize scarce R&D resources. So far Japan seems the first country where even BR is led by the industry. In a certain value system the free-market system is considered as ideal. Then BR done by the market principle is to be more highly evaluated than that by the ivory tower, the governmental or military power though the academism or militarism value system evaluates differently.

Industrialism may be more acceptable to the public today than academism or militarism but welfarism may be more acceptable than industrialism. Besides the Large-Scale Projects stated in the subsection 4.3, MITI has also provided the Program of R&D on Medical & Welfare Equipment Technology since 1976 (Tale 6.23) which followed a similar program of STA in the early 1970's. Medical or welfare R&D may be expected to have a strong side-effect on industrial technology through bio-technology, new materials, electronics and informatics. As one of its characteristics which is different from the industrial R&D promotion discussed in the subsection 4.3, the products of medical and welfare R&D can directly be purchased by local and central governments through public hospitals and welfare offices and indirectly recommended to the users through the public health insurance program after which the private ones are expected to follow [Eto 1980C]. But the procurement and recommendation activities are impossible for MITI which administers neither hospitals nor health insurance programs and these activities can be made only by the Ministry of Welfare. As the Ministry of Welfare administers as many as ten research institutes and centers (Figures 2.1 and 2.2), the only remaining problem is its planning functin. Unlike MITI which has AIST as the unified single organ to administer the whole R&D activities in MITI, the Ministry of Welfare lacks such a unified organ. An informal mechanism may be needed for the Ministry of Welfare to lead the national R&D activities.

Table 6.23 R&D on Medial and Welfare Equipmet Technology

R&D that has been completed up to 1981

(in million yen)

R&D theme	Cumulative budget
1. Medical equipment technolgoy	2,117
(1) Multichannel automatic biochemical analyzer	247
(2) Automatic differential blood cell analyzer	264
(3) Artificial heart for clinical use	476
(4) Portable artificial kidneys	596
(5) Laser scalpel	534
2. Welfare equipment technology	623
(1) Modular-type electric wheelchair	223
(2) Braille reproduction device	141
(3) Rehabilitation devices for the physically handicapped	159
(4) Multifunction bed	100
Total	2,740

(to be continued to the next page)

Table 6.23 (cont'd)

Ongoing R&D in 1982

(in million yen)

R&D theme	Total budget up to FY1981	Budget for FY1982
1. Medical equipment technology		
(1) Positron CT	355	115
(2) Liver function support equipment	364	109
(3) Computer-assisted polysomnography and drug monitoring	30	146
2. Welfare equipment technology		
(1) Implanted artifical tympanum	378	51
(2) Guide device for blind people	213	83
(3) Vocal and speech training device for people with impaired speaking ability	204	88
(4) Leg-function assiting equipment	156	197
(a) Motorized artifical legs	135	100
(b) Electric wheelchair with movable seat	21	97
(5) Book-reader for the blind	-	34
Total	2,234	823

(Source: [AIST a])

The recent progress in life science has been strongly motivated by the military purpose for biological weapons. But Japan is one of the few nations in a position to declare a pure welfare technology system in opposition to the weapon-oriented one.

Pastoralism or agriculturism may be as well acceptable as welfarism especially in context of environmentalism and internationalism. The raise of food productivity is urgently needed to save the mankind from the existing hunger in the third world, and ruralization is desired as a counter-measure for the environmental protection in certain circumstances. As Japan was an agriculture-based nation only till the 1950's, she still keeps accumulated knowhows about plants, insects, fish, leavening and fermentation (i.e., sake brewing; soy-beans and rice malt for condiments; fermented soy-beans for vegetable cheese, milk and curds; fish guts and so on) and gardening moss among eldery farmers. To create jobs for eldery farmers in rural areas is important to prevent excessive migration into urban areas, and to put the knowledge of these farmers to creative use is important as a part of welfare programs besides their expected proper effect on R&D itself. To collect their knowledge into data base before its loss may be useful in life science or bio-technology which need huge knowlege on several trillion species. In addition, this may stimulate the advance of information technology for data base or so-called knowledge engineering. Having emerged just recently from an agricultural country to an industrialized one, Japan is in a very favourable condition to recall and collect such knowledge.

The basic policy of the Ministry of Agriculture has been to protect the farmers existing way of life and hence the existing mode of agriculture against a national policy to industrialize the whole nation and to import the US agricultural products since MSA (Mutual Security Agreement in 1953) which exchanged the military defence of Japan with the economic defence of US agriculture. In these three decades, the Ministry of Agriculture has suffered defeats to the industrial policy which works to acquire the US market in exchange for US food imports. After the repeated defeats, the agricultural R&D may be able to lead the industrial technology in the upcoming age of bio-technology. The Ministry of Agriculture administers as many as fifteen research institutes, centers and experiment stations plus four research institutes and laboratory under its jurisdiction and hence has a sufficiently large reserach staff. Moreover, the Ministry of Agriculture has the Agriculture, Forestry & Fisheries Research Council as a single unified, though not administrative but counseling, organ to guide these research organizations (Figures 2.1 and 2.2). Thus it may have certain planning powers from the standpoint of organizational structure. Similar to welfare technology, agriculture technology can also be closely and naturally linked with the governmental policy in that the governments subsidize the agriculture in many countries. This will make the grant to agricultural R&D more internationally and domestically acceptable than that to industrial R&D. In addition, this will contribute to the nation-wide diffusion of scientific knowledge to every corner of the land and stimulate science education in rural areas. These effects will bring about scientifically educated manpower all over the country in the next generation and will multiply future manpower in number and quality.

7. CONCLUSION

Japan is known as a nation of consensus. But the consensus lacks and political conflict even exists concerning R&D policy. The process of persistent political conflict has generated a overlapping structure of the various organs for science and technology policy each of which was given an ad hoc purpose for some particular political reason. This historical condition makes the formal mechanism for policy formation quite unsystematic.

Instead, the informal mechanisms have been developed in the forms of informational control. Among various forms, the most remarkable ones are the nation-wide technology forecasting and the symbol operation in terms of national projects. The responsibility averse behaviour in the industry associated with the Japanese bottom-up decision system provides the ground for the industry willingly to accept governmental guidance for the long-range technology development strategy. The governmental activities serve as the decision aid to the industry. These informal mechanisms work together to coordinate R&D activities across firms and to unify separated R&D activities into the systematized national stream.

Finally, the international criticism that Japan is still a follower of Western technology is refuted with evidences. But a careful understanding on the Western feeling behind the criticism is attempted and the original idea implicitly existing behind Japanese technology is identified as an industry-pulled technology system in contrast to the traditional academism-pushed or militarism-pulled ones in Western culture.

REFERENCES

Abe, Y., H. Inose, T. Nishikawa, K. Shimo and A. Tezuka, 1982 : Policies and practices of government support for basic and applied science, in A. Gerstenfeld (ed), Science policy perspectives ; USA-Japan, 125-174, Academic Press, New York, USA.

Agency of Industrial Science and Technology (AIST), 1982: AIST (English edition), (abbreviated as AIST a), AIST, the Ministry of International Trade and Industry (MITI), Tokyo, Japan.

Agency of Industrial Science and Technology (AIST), 1982 : Trends of principal indices on research and development activities in Japan (English edition), (abbreviated as AIST t), AIST, The Ministry of International Trade and Industry (MITI), Tokyo, Japan

Bradbury, F.R., B.G.Bordroghy, F.J.P.Clarke and P.Jervis, 1981 : European Economic Community industrial research priorities for the early 1980s, R&D Management 11, 157-164

Bureau of Statistics, the Prime Minister's Office : Report on the survey of research and development, The Printing Office, Tokyo, Japan (annual report)

Burke, C.E. and D.S.Price, 1981 : The distribution of citations from nation to nation on a field by field basis - a computer calculation of the parameters, Scientometrics 3, 363-378

Carpenter, M.P. and F.Narin, 1980 : The subject composition of the world's scientific journals, Scientometrics 2, 53-63

Dajani, J.S., M.Z.Sincoff and W.K.Talley, 1979 : Stability and agreement criteria for the termination of Delphi studies, Technological Forecasting. 13, 83-90

Dalkey, N.C., 1968 : Experiments in group prediction, RAND P3820, RAND Corporation, Santa Monica, Calif., USA

Dalkey, N.C., 1969 : The Delphi method : an experimental study of group opinion, RAND RM5888PR, RAND Corporation, Santa Monica, Calif., USA

Dobrov, G.M., 1977 : Systems assessment of new technology in decision-making in government and industry, IIASA WP-77-8, International Institute for Applied Systems Analysis, Laxenburg, Austria

Dobrov, G.M., 1979 : A strategy for organized technology, Technological Forecasting. 13, 257-271

Eto, H., 1972 : Methodological discussion on the Delphi method, in Y.Hayashi (ed), Basic design of Japanese R&D system, Institute of Future Technology, Tokyo, Japan

Eto, H., 1979A : Intercorporate technology transfer in Japanese industry, Interfaces 9, (2), 70-71

Eto, H., 1979B : Relationship between basic and improvement innovation - Development of industrial policy of Japan, Presented at IIASA workshop on innovation policy and firm strategy, Laxenburg, Austria, 4-6 Dec.

Eto, H., 1980A : Problems and lessons of Japanese technology policy, R&D Management 10, 49-59

Eto, H., 1980B : Characteristics of Japanese innovation (translated in Hungarian), Proceedings of Hungarian Academy of Sciences 20, 313-327

Eto, H., 1980C : The recent trend of Japanese R&D management and its prospect of industrial R&D strategy, Presented at international meeting on industrial R&D strategy and management - a challenge for the 1980s, Manchester, UK, 29 June-1 July

Eto, H., 1982 : Preference of technological infrastructure to superstructure in the Japanese industry, Presented at the scientific meeting on theory and experience on effective use of scientific results in economy, Novosivirsk, USSR, May

Haigh, M.J., 1982 : Citation analysis of foreign sources in Japanese geographic serials, Scientometrics 4, 195-203

Hayashi, Y. (ed), 1972 : Basic design of Japanese R&D system, Institute of Future Technology, Tokyo, Japan

Hill, K.Q. and J.Fowles, 1975 : The methodological worth of the Delphi forecasting technique, Technological Forecasting 7, 179-192

Inhaber, H. and M.Alvo, 1978 : World science as an input-output system, Scientometrics 1, 43-64

Inose, H., T.Nishikawa and M.Uenohara, 1982 : Cooperation between universities and industries in basic and applied science, in A.Gerstenfeld (ed), Science policy perspectives : USA-Japan, 43-62, Academic Press, New York, USA

Ito, D., 1980 : Analysis of the Japanese contemporary bureaucracy (in Japanese), The University of Tokyo Press, Tokyo, Japan

Murray, T.J., 1979 : Delphi methodologies : a review and critique, Urban Systems 4, 153-158

Ministry of Education (ME) : Indicators of education (in Japanese), The Printing Office, Tokyo, Japan, (annual report)

Nelson, B.W., 1978 : Statistical manipulation of Delphi statements; its success and effects on convergence and stability, Technological Forecasting. 12, 41-60

Price, D. de S., 1980 : Comments on "U.S. science in an international perspective," Scientometrics 2, 423-428

Rieker, W.S. and S.J. Sullivan, 1982 : QC circles' trend in the US, a concern, ASQC Quality Congress Transactions, Detroit, USA, (Japanese translation, Engineers, No.412, 14-19, 1982)

Rosenberg, N., 1980: Comments on "indicators of the impact of R&D on the economy," Scientometrics 2, 387-393

Science Council, 1981 : The Science Council of Japan, outline (English edition), The Science Council of Japan, Tokyo, Japan

Science and Technology Agency (STA), the Prime Minister's Office (abbrevaited as STA i) : Indicators of science and technology (in Japanese with English explanation), The Printing Office, Tokyo, Japan, (annual report)

Science and Technology Agency (STA), the Prime Minister's Office, (abbreviated as STA p) : Report of survey of R&D in private firms (in Japanese), STA, Tokyo, Japan (nearly annual report)

Science and Technology Agency (STA), the Prime Minister's Office, 1982 (abbreviated as STA tf) : Report of technology forecasting (in Japanese), STA, Tokyo, Japan

Science and Technology Agency (STA), the Prime Minister's Office, (abbreviated as STA sta) : STA - its roles and activities (English edition), STA, Tokyo, Japan, (annual booklet)

Science and Technology Agency (STA), the Prime Minister's Office, (abbreviated as STA w) : White Paper on science and technology (in Japanese occasioanly with English abstracts), The Printing Office, Tokyo, Japan, (annual paper)

Shirane, R., 1973 : Technology assessment (in Japanese), Nippon Keizai Press, Tokyo, Japan

Soetes, L.G. and S.M.E. Wyatt, 1983 : The use of foreign patenting as an internationally comparable science and technology indicator, Scientometrics 5, 31-54

Straszak, A. (ed), 1981 : The Shinkansen program; transportation, railway, environmental, regional and national development issue, IIASA CP-81-S2, International Institute for Applied Systems Analysis, Laxenburg, Austria

Tsuda, M., 1977 : Logic of the Japanese management (in Japanese), Chuo Keizai Publishing C., Tokyo, Japan

Tsuji, K., 1969 : Study of the Japanese bureaucracy, new ed., (in Japanese), The Universtiy of Tokyo Press, Tokyo, Japan

Tsukahara, S. and K. Yamada, 1982 : A note on time lag between life cycle of discipline and resources allocation, Research Policy 11, 133-140

Zsindely S., A. Schubert and T. Braun, 1982 : Editorial gatekeeping patterns in international science journals, Scientometrics 4, 57-68

R & D Management Systems in Japanese Industry
H. Eto and K. Matsui (editors)
© Elsevier Science Publishers B.V. (North-Holland), 1984

Chapter VIII SOFT SCIENCE IN JAPAN
--- TEN YEARS' EXPERIENCE IN MITSUBISHI ELECTRIC CORPORATION ---

KAZUO YANAGISHITA

Soft Science Group
Mitsubishi Electric Corporation
1-1, Tsukaguchi-Honmachi 8-chome
Amagasaki, Hyogo 661, Japan

Herein is described the experience of soft science activities in Mitsubishi Electric Corporation and Japan in this decade which enscopes the various fields related to technology development policy.

1. THE DECENNIAL ACTIVITY OF SOFT SCIENCE GROUP IN MITSUBISHI ELECTRIC CORPORATION

1.1 Soft Science

The word "Soft Science" first appeared in the Science and Technology White Paper published by the Science and Technology Agency of Japan in 1971. It is defined like this; Soft science is a new management method of forecasting, analyzing, planning controlling and evaluating to solve complex problems with various entangled factors.

As the word is made-in-Japan English, it was hard to be understood internationally, but it is becoming popular recently for instance there held an seminary "The Status of the Art of Soft Science in the Battelle Memorial Institute" in Tokyo in 1978.

We can understand mechanical engineering, electrical engineering etc. which are connected with hardware as "hard science", and also we can understand technology forecasting, demenand forecasting, factor analysis, multivariate analysis, PERT, research and development administration, PATTERN etc. which are linked with software as "soft science".

Systems engineering and policy science have similarity with soft science in some points.

1.2 Social background to bring forth soft science

Let's consider the reason why soft science was born from the viewpoint of social background.

The first reason is social phenomena became complex and value system became diversified. Therefore, it is recognized that traditional hard science is not sufficient. For example, Tokyo International Airport at Narita was completed as hardware but it could not be in use for almost ten years due to lack of software like compensation for inhabitants nearby.

This is the confrontation of value system between pro-airport party who believe civilization like having international airport is good and anti-airport party who want to live on farms and protect environment.

Another example is a railroad. It was a epochmaking event that a railroad was constructed in countryside before the War. It brought convenience in travel, lessened time to go to city and invited introduction of culture. It was a very happy event and on the opening day all the pupils of the primary school welcomed the first train waving national flags on the platform and sent bouquet to the operator. And the master of the station became Very Important Person (VIP) of the village like village headman, principal of the primary school, postmaster and superior of the Buddhist temple.

Besides, the construction could be directed by civil engineer, mechanical engineer and electrical engineer.

Everything is opposite at presnt. New railroad is deemed to be the motive force to bring public hazard such as noise and vibration and destroy rural beauty by housing development. Therefore, on the opening day university students may wave iron bars instead of national flags or may throw flame bottles instead of bouquet. To prevent such accident, it is necessary to measure noise and vibration along the railroad and to compensate the decrease of egg production for poultry farms and provide air conditioners for houses with unopenable windows due to vibration and noise.

Such jobs can not be done by only civil, mechanical and electrical engineers but also agricultural and architectural engineers.

The second reason is higher speed and wider range of technological change which used to be slow and narrow.

In 1796, James Watt invented steam engine and it brought about industrial revolution in England by 1840's. The engine was sent to Japan and caused revolution in the country by the end of the 19th century. The technological change in those days are in such slow speed.

But now technological change can propagate much faster and broader owing to international interchange of information and people with the development of communication and transportation. For example, transistor was invented by W. Shockley, J. Bardeen and W. H. Brattain at Bell Telephone Laboratory in 1948, and three years later it was applied to radio by Sony in Japan, and transistor radios were mass produced and sold all over the world in several years later.

Thus high speed and wider range change of technology needs appropriate forecasting and planning.

The third reason is the difficulty in grasping the situation and forecasting the future which delayed response of people and society. For example, economical growth of Japan is forecasted by professors, economists, bankers, analysts and executives in newspapers on January the 1st. The forecast differs much among them from minus growth to plus two figure percent.

The difference in forecast for only one year by the intelligent people shows the difficulty of forecast for the future. Consequently it is very hard to respond to such present situation and the future. One example was seen in 1960's in Japan when economic growth was high and those who work in firms were happy but pensioners and mother and children families were hit by galloping inflation because the increase of pension and livelihood assistance was delayed and unfairness among Japanese citizens was expanded.

The fourth reason is the development of the tendency to respond to the problem scientifically and by neither intuition nor experience.

The late Mr. Hubert, H. Humphrey, former vice-president of the United States once told, "It is very difficult to consider 10 or 20 or more factors at a time when a man has to make decision. In the space age of present, man must consider thousands to hundred thousands factors simultaneously. Human capability does not change in the past twenty years, but complexity of decision making in the government and firms was multiplied more than one thousand times. Therefore, it is urgently necessary to solve decision making rationally and a new method must be developed."

Thus intuition and experience became useless in the fast changing society.

1.3 Objects of soft science

As it is evident from the definition, the objects of soft science are big problems with complicated factors which may be called policy issues.

An example is public hazard which is a big problem related with technical issue at its source, legal issue of the competent authorities, geographical and meteorological issue of affected area, medical and physiology issue of the sufferers, physical, chemical and engineering issue of prevention etc.

Environment problem is related with natural ecology of animals and plants, aesthetic sence, countermeasures against calamities like landslide, right to accept sunshine, right to command a fine view, legal issue of natural environment protection law etc.

Resource problem must count geological abundance of resources, resource cartel strategy of underdeveloping countries, save resource efforts in industrialized countries, development of alternative materials, progress of recycling technologies etc.

Urban problem includes population issue, policy of firms, traffic jam, progress of communication, centralization of information, air pollution, noise, crime, housing, garbage treatment, energy, physical distribution etc.

Traffic problem contains differentiation of start time of each firm, government and public office, school and store, city planning, traffic control, technological progress in traffic, transportation cost, popularization of automobiles, public transportation, traffic accidents, exhaust gas from automobiles etc.

Welfare problem affiliates increasing aged people, protection of the handicapped who are the weak of the society, medical care sufficient for necessary patients, tax increase to support welfare, training of caseworkers and volunteers etc.

Education has so many problems, for instance, what is the suitable distribution of graduates to meet future structure of Japanese industries, revolution of education system, cost of education, ordeal of examination, too many private preparatory schools, education with sufficient time, deviation of text book, the school district system, invasion into other school district, unsuccessful examinees, obtaining admission to university by unfair means, violation in school, retarded students, the cramming system etc.

Calamity prevention problem contains revision of fire service act, foretelling an earthquake, change of building materials, progress in calamity prevention technologies and equipments, panic psychology, refuge training, emergency rations, rescue system etc.

Crime prevention problem deals with trends of criminals, progress of crime investigation technology, training of policemen, crime prevention educaion, the rehabilitation of offenders etc.

Protection of consumers treats revision of the commercial law, consumer education, Fair Trade Commission, consumers' organization, commodity check system, study of domestic science, home economist in business (HEIB) etc.

The objects of soft science are large scale and complicated problems as above.

1.4 Characteristics of soft science

Soft science has several characteristics. The first characteristic is intuitive. It may sound contradictory to scientific but it is often found in the newest science for there is no authorized theory established. Vast experiments like thoughtless carpet bombing are impossible due to limited R & D personnel, time and budget. Then several hypotheses are suggested and they are verified by limited experiments. Intuition plays an important roll in building up the hypotheses.

The intuition is also effective to final clues to problems in soft science.

The second characteristic is normative. The word norm implies a standard to follow in evaluation, decision and action. Normative approach in soft science first evaluates future goal, necessity, need or mission and then decides what should be done from now till then.

For example, in public hazard problem maximum permissible concentiation of toxic gas in air is defined to be X ppm to minimize bad influence for human health. And toxic gas concentration from automobiles is defined to be Y ppm and from steam generation plants Z ppm etc. so that the sum of them may not surpass X ppm. It is not our approach that automobile exhaust contains S ppm of toxic gas and steam generation plant does T ppm etc. therefore, town air contains S+T+..... ppm of toxic gas.

The third characteristic is interdisciplinary. The objects of soft science are big and complicated and can not often be solved by single discipline but can not be solved by dividing into several disciplines, but solved by cooperation of plural disciplines.

Interdisciplinary area such as medical electronics, solid state physics, computer aided instruction, social engineering etc. are fast progressing area.

The fourth characteristic is system approach. It has objective, budget and dead line.

For instance, a young men's group is going to hold a dane party on Saint Valentine's day and want to buy stereo audio set which consists of record player, amplifier, speakers and disks. If a man insists to buy expensive speakers to make high fidility sound, then other components become relatively of low class and consequently poor sound will be heard. A man wants the best quality components and budget will be doubled. A man tries to make the set by himself to get the best sound at lowest cost and it will be made in March. Such mistakes will be avoided by system approach in soft science.

The fifth characteristic is problem solving. Of course all the problems can not be solved. But soft scientists try to solve them, and even if the best solution can not be found, they present the second best solution. They are different from irresponsible critics.

The sixth characteristic is ambiguity. This is also contradictory to be scientific. Questionnaire, which is often used in soft science research, is answered with fifteen percent of incorrect answers statistically. In making this fifteen percent correct, questionnaire may become insignificant.

For instance, in public opinion survey for one month prior to the election is interesting but sometimes incorrect because fifteen percent of people change their opinion after they saw the survey result. The correct method is to survey when they come out of voting booths which is too late and nonsense as prediction.

There is a principle of uncertainty in modern physics which tells when physical values are measured some ambiguity exists, for instance, if position of an electron is measured correctly its time of existence at the position is uncertain at some allowance. By the introduction quantum physics made great progress. This shows that we can manage ambiguity if we know the allowance.

The seventh characteristic is future oriented. The objects of soft science often deal with future problems. Of course we study past history of a subject, but our interet is different from historians and archaeologists. We study the history to predict the future trend.

1.5 Requirements for soft scientists

There are three types of soft scientists, namely, surveyor, analyst and consultant, and a man can serve as two or three types.

A) The requirements for surveyor are as follows. At first he must have profound knowledge and experience for information sources and be able to judge correctly where to go to obtain necessary information.

Secondly, he must have wide human relation because the newest information will be obtainable from human mouth. Newspaper tells as news of yesterday, weekly magazine shows us what happened in last week, monthly magazine informs us events of last month and book brings us knowledge of last year. Information told by a specialist in some field is the information of now. Therefore, a surveyor who has wider human relation can get better information.

Thirdly, he must master interviewing technique. Both interviewer and interviewee are busy men in most cases. The technique is essential if he wants to know everything he wants to know in a limited time. Questionnaire can be sent to some interviewees beforehand and get answer one by one but can not be sent to other interviewees for some reasons. There are some difficulties to make questionnaire beforehand.

Some interviewee speaks his theory eloquently irrespective of the interest of the interviewer. Other interviewee speaks with an air of importance to teach general knowledge which the interviewer already knows. The interviewer should make him return to the main subject in a polite manner lest he should waste time.

Fourthly, he must have good understanding. But it is difficult to understand all the points because he went to hear what he did not know. He must learn about the problem before the interview by reading introduction books or basic knowledge book for the problem but borrowed plumes are sure to come off sooner or later. Good solution for such situation is to go with a person with an elementary knowledge for the problem to understand the interviewee's answers although he can not answer by himself. If he ask too primitive question, the interviewee will despise and not answer main question of importance. For instance if he go to survey super conductive alternating current generator and ask "What is alternating current?" the interviewee will loose intention to explain super conductive generator. On the contrary, if he understand too well and find fault in contradictory expression, the interviewee will be hurt. Interviewer is no more police detective than interviewee is suspected person.

Fifthly, he must be curious. If he is always curious he may get unexpected by-products. Interviewee sometimes talk different subject from the main purpose of the interview, which is very valuable information. If he ignores it he will regret when he return home.

B) The requirements for analyst are as follows. At first he must master some of soft technologies such as Delphi method, scenario writing, trends extrapolation, relevance tree method, PATTERN (Planning Assistance Through Technical Evaluation of Relevance Number), morphological analysis, cross impact matrix analysis, learning curve, experience curve, PPM (Product Portofolio Management), PERT(Program Evaluation and Review Technique), CPM (Critical Path Method), envelope curve, inter-industry economics, brainstorming, brain-writing, science fiction etc.

Secondly, he must have analysing ability. Only genius can correctly grasp the whole aspect of things by intuition, and ordinary man must seize things analytically by dividing them into simple elements.

Analysis is inevitable to improve present status with minimum cost. For instance when an automobile does not start, there is no need to buy new car but to replace battery is good enough which was found by analysis.

Thirdly, he must have analysing ability for statistic data which is only an array of figures for outsiders, but if it is well analyzed it is rich in contents.

Fourthly, he must have logicality. If there is no gap in logic it will increase persuasive power. If there is one gap in logic, the content is greatly discounted even though the rest 99% are accurate in logic, and lose his credit completely.

C) The requirements for consultant are as follows. At first he must have problem finding ability. Client may bring in problem saying "This is the problem, please solve it" but sometimes it is not correct if objectively judged.

For example, a patient come to a doctor and told "I have a pain in the abdomen, please operate my appendix." But if he made a correct diagnosis he found it peritonitis instead of appendicitis.

It is very important for a consultant to listen to his client carefully and to judge what is correct problem.

Secondly, he must have problem solving power. If he can not solve the problem it is of no use to find correct problem. The doctor is unreliable if he said "This is not appendicitis but peritonitis, which I can not operate."

Thirdly, he must have persuading power. When he wrote a good report, if client said "Thank you very much. We will keep it for future reference." he can get consultation fee for this time but can not expect next order. Consultants' job is not to increase clients' knowledge but profit.

1.6 Establishment of Soft Science Group in Mitsubishi Electric Corporation

(1) Setting research themes

Let's explain how Soft Science Group (hereafter SSG) is established in Mitsubishi Electric Corporation (hereafter Melco). There were two ways to set new research themes in Melco. One is requested research which is requested by head office, factories or sales offices when our customer wanted a product which Melco is not producing yet. Our laboratory examine the requests from the point of personnel, facility and budget. And some of them are accepted as request research. This is the first route.

The second route is called voluntary research. Researcher gets an idea by himself, by reading magazine, by attending academic meeting, by visiting uninversities etc. and he wants to do the research. If his boss agrees he can do the research. This is the second route.

Both routes worked well and no trouble happend till 1969 when Melco adopted new slogan of "Advanced and ever advancing Mitsubishi Electric" instead of old "Service through quality" since its foundation. In later 1960's future industry looks very prosperous. Futurist like Herman Kahn inspired us with rose-coloured future industries such as atomic energy, information industry, ocean development, space development etc.

To enter these future industries, the two routes of setting research themes above mentioned were found insufficient.

One example will explain how they are insufficient. Food problem is very important for Japan because she imports more than 60% of her food. In such environment how can Melco set new research theme of industrialized food production system, that is, to make vegetable in a factory?

Those customer who want to industrialize food production system rarely visit Melco but they go to agricultural machinery manufacturers to order the new system.

Therefore, the first route of requested research does not work well.

Is the second route of voluntary research effective, then? Unfortunately, there is no agricultural researcher who reads "The industrialized food production system" on an agriculture journal. Consequently, Melco has nothing to do with food crisis.

The SSC was established to catch social change and to survey and plan future research theme in June 1973. It was a group in Melco's Central Research Laboratory but it found not only future research themes for the laboratory but development targets for factories during its activity. Then SSG was transferred to headquarters, development department in 1976 in organization but is still staying in the laboratory and surveying new research themes and market need for new products.

(2) Tasks of SSG

The SSG has five main tasks of technology forecasting, demand forecasting, new products planning, management strategy planning and marketing.

The first task is technology forecasting. Research resources such as personnel, budget and facility are limited and effective use is essential. For this purpose more resources should be allocated for projects which will bring forth fruits in immediate future rather than distant future. It is a big question that which technology will be realized sooner. Therefore, technology forecasting is very important.

For instance, there are two alternatives of fast breeder and nuclear fusion in energy research. If fusion is realized faster there is no need to research fast breeder. But technology forecasting tells us that fusion will be actualize in the twenty-first century. Then fast breeder research is necessary in order to save uranium for uranium will be spent all by light water reactor.

Such task is done by SSG to forecast which technology will appear first and to help manager to select research themes, allocation of researchers and budget.

The second task of SSG is demand forecasting. The reason why SSG forecasts demand is to select research themes from market size, that is, if expected time of realization and necessary research resources are same, laboratory manager of business firms should choose the research themes with larger market.

The demand forecast of television sets and motors are not SSG's task but sales offices' and factories'. The demand forecast of SSG is to estimate influence of research on market before starting the research. What most researchers lack is this market orientation or need orientation as they are strongly technology oriented. Hence we have a lot of clients in Melco when they wanted to research soler energy, anti-pollution apparatus, personal computer, medical electronics etc.

The third task of SSG is new product planning. This is being watched by factory managers with keen interest. We study by ourselves as well as with engineers sent to SSG from factories as trainees. Melco has training system among labortories and headquarters, factories and sales offices. We had eighteen trainees two from headquarters and the rest from factories. An example of their activity to find a new product will be shown in section 2 of this chapter.

The fourth task is business strategy planning for new businesses. Strategic plannings for conventional businesses and extraporation of them are planned by respective divisions and president office.

Our planning is to locate new product among its surroundings, in other words, to draw map around new product. For example, when we plan a new product of medical electronics we survey medical industry as a whole, because only concept of new product can hardly excite factory manager. It is important to segment market for the new product, to differentiate from competitors' products and to suggest next product.

The fifth task is marketing. We do market survey to forecast demand, to plan new product and to find new application of R & D fruits.

Sometimes we go abroad to survey market needs, for instance middle east where Melco exports heavy electric systems such as transformer substations. This business will be situated soon for the population is not so big and industry is not well developed in middle east and life of substation is too long.

Then the manager of export division asked us to find new projects in middle east to draw oil dollers.

2. NEW PRODUCT PLANNING BY SOFT SCIENCE GROUP

One activity of the soft science group (hereafter SSG) is new product planning which is expected most by factories.

Manager of the application apparatus works in heavy electric machinery division asked us to find new product. Three products were manufactured in his works, namely electric furnace, induction heater and electric welder.

They are supplied to iron and steel industry and automobile industry. Both industries fluctuate simultaneously so that workers in the works must work overtime when business is brisk and they have to stay at home when business is dull. The manager, therefore, wanted the fourth product to make his works stable.

His requirements for the new product are as follows.

(1) It should have different response for business fluctuation with the three conventional products.
(2) Its major components should be manufactured in the works.
(3) It should be a plant rather than mass production product.
(4) It should cost several ten million yen.
(5) It should have five billion yen market in three years.
(6) It should have no competition among Mitsubishi Group.

The requirements need some explanation.

(1) If the fourth product has different responce, the operation of the works will become more stable.
(2) If he can make more components rather than buy, he can give more job for his workers.
(3) The works has many plant engineers but no mass production lines.
(4) If it is several hundred million yen, it is too big for the works and they cannot hundle when two orders come at the same time. If it is several million yen, it is too small for the salesmen to visit customers.
(5) If the market is such size he can sell even if competitors appear.
(6) Some companies in Mitsubishi Group is quite competitive with the works. The manager does not like such competition.

He sent a trainee ot ten years experience to SSG.

The SSG surveyed in the following process.

(1) Need survey in works

Melco is producing more than 30,000 kinds of products. It consists a small industrial world. Something it needs may be needed by Japanese industry. This is the main reason why we visited more than twenty works in Melco. Other reasons are easiness to get appointment and to prevent secrecy of survey.

Before we visit works we sent questionnaire to ask troubles which the works may be suffering in automation, labor saving, working environment, energy saving, water industrial waste, disaster prevention, crime prevention etc. Then we visited and investigated on sites.

(2) Seed survey in laboratories

We visited research group leaders in Central Research Laboratory and Products Development Laboratory of Melco to find technological seeds for new products.

(3) Technoloy survey in the works

The manager of the works wanted to make most of the components of new product instead of buying. It is important to know what kind of component can be produced by the works. We asked all the engineers in the works to fill in blanks to show his speciality, experience and present level. These reports were analyzed and technological capability of the works became clear.

(4) Survey of business fluctuation

The aim of this survey is to find products with different fluctuation from the three products of the works. We analyzed business fluctuation and shipping records of industrial product for thirty years. And found several products sold quite different from the three conventional products. They were civil engineering machinery, anti-pollution equipment, agricultural equipment, disaster and crime prevention instruments food and confectionary equipment and leisure goods.

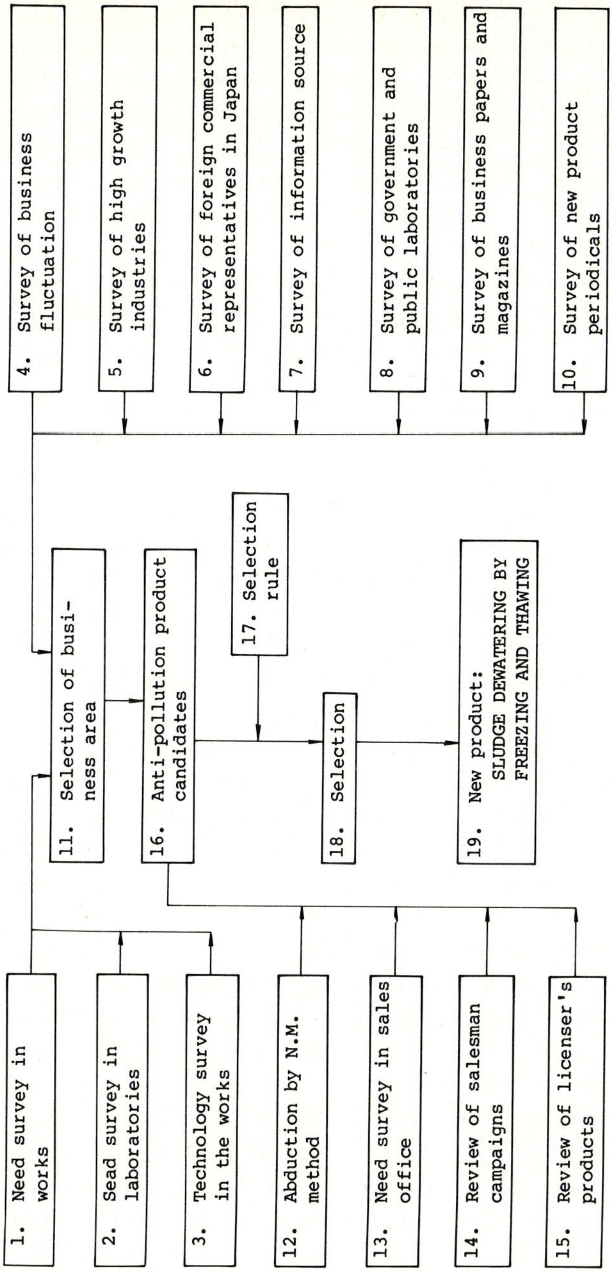

Fig. 1 New Product Planning for Application Apparatus Works

(5) Survey of high growth industries
We read many books on futurology, attended many seminaries for new product planning such as vegetable factory, antipollution apparatus etc.

(6) Survey of foreign commercial representatives in Japan
We visited U.S. Trade Center, British Trade Center, West Germany Chamber of Commmerce and Industry in Japan to survey industrial activities especially new product licensing in those countries.

(7) Survey of Information Sources
We visited catalog houses where they sold catalogs of new products. We also visited a general trading company to survey license offers.

(8) Survey of governmental and public laboratories
We visited several governmental and public laboratories to find new technological seeds which private laboratories did not research.

(9) Survey of business papers and magazines
We read several business newspapers, economical and industrial ones, to catch economical and technological trends and also read more than thirty business magazines for the same purpose.

(10) Survey of new product periodicals
There were a few periodicals specialized in introducing new products, which we read to know the intension of other companies to plan them.

(11) Selection of business area
From the information obtained by the above process we decided to develop anti-pollution equipment. In decision we considered four factors of market, technology, production and sales. The market factor consisted of market size, growth rate, elasticity for business fluctuation, life cycle, influence by politics, safety, non-pollution and social acceptability. The technology factor covered familiarity with the works, strength in patents, backup

by laboratories, expectancy of new products and possibility to make product series. The production and sales factor included type of order, type of production, percentage of hardware, sales route and competition within Mitsubishi Group.

(12) Abduction by N.M. method

The N.M. method is a Japanese method to solve problem, that is a kind of creative engineering developed by Mr. Nakayama, Masakazu.

We generated many ideas to treat industrial wastes.

(13) Need survey in sales office

We went to our sales office to survey what kind of needs they had from their customers in pollution troubles.

(14) Review of salesman campains

Sometimes we had in our company sales companies to find new products. We reviewed them from the view point of anti-pollution equipments.

(15) Review of licensers products

One of our lincenser covered very wide range of products, for some of which we were licensee. If we can find another product to be licensed, the negociation must be easier than a new licenser. Therefore, we checked catalogs and other documents to find candidates for licensing.

(16) Anti-pollution product candidates

From the survey (1) to (15), we found eleven candidates of new product for the works from anti-pollution field.

(17) Selection rule

Before we select one from eleven candidates, we made selection rule, which consisted of six factors of research, technology, development, market, sales and competition. The research factor means whether research related to the candidates is underway in Melco laboratories or not. The technology factor implied whether the works or Melco laboratories had enough technology to cope with the candidates. The development factor covered the difficulty in the development, necessary manpower, time, facilities and cost for the development. The market and sales factors counted market size, what kind of industries would be our customers, expected market price, percentage of added value by the works, response to business fluctuation, life cycle and profitability. The competition factor involved possibility of licensing, who were competitors and who had patents.

(18) Selection

The selection was made by the selection committee which was consisted by the manager and his staffs, SSG members and the trainee. The selection was done by filling blanks with five grade figures. Then figures in each blank of all the members of the committee were added and the candidates with the best score was selected as new product.

It was sludge dewatering by freezing and thawing. Sludge is a mixture of solid suspended in water, which is very hard to remove from sludge. Therefore, sludge generating works must pay big amount of money for sludge disposal companies which dispose sludge by weight.

There were many kind of sludge; plating sludge, chemical etching sludge from electrolysis processing, printed circuit boards, semiconductors etc., red mud sludge from aluminum factory, cheese whey, sewerage sludge, etc.

When some kind of sludge are freezed, suspended solid particles form flocks of solid. And then it is thawed the flocks sink and make deposit which is very easy to dewater. Sludge with 99% of water can be dewatered till 50%. This means voluime and weight is reduced to 2% of the sludge. It can save payment for sludge disposal company to great extent.

After two years of our survey, it was commercialized and has been sold ever since. The main customers are municipal offices who treat sewerage sludge.

3. LONG RANGE R & D SKILLS DEVELOPMENT MODEL

3.1 Purpose

3.1.1 Purpose of the study

Research and development (hereafter R & D) activity in Japanese companies, where the practice of life-long employment is common, inevitably presents such problems as the constant and proper development of competent researchers.

Faced with an economy with a low growth rate, a new strategy for R & D should be explored and should be different from that in the past when the growth rate of the economy was high. Since personnel cost occupies the biggest expenditure of R & D activities today, it is impossible to employ anabundant number of researchers in R & D activities. In order to effectively institute R & D activities within an enterprise, therefore, it is necessary to utilize the rather limited number of researchers in a company to realize the highest productivity.

While we employ to company-wide skills inventory system, the system as such is not able to meet the requiremnts of R & D activities in terms of locating suitable personnel for a given task. Hence, researchers have often to be looked for from among the concerned parties only after a problem has become explicit. This procedure certainly is not the best way to utilize all of the researchers in the corporation.

In this paper we are going to offer a conceptual framework in which effective utilization of researchers in an R & D laboratory can be achieved. The framework aims to provide researchers with as many chances to work as specialists as possible in the Japanese climate of life-long employment.

3.1.2 The Method of the Study

In the conceptual framework for long range planning for R & D skills development, we assumed 216 situations which an R & D manager has to consider in order to plan R & D skills development to cope with a new R & D theme. Further, in order to fill the gap between the technological potential of his unit and the required ability to conduct the research in the area, he must choose one out of four alternatives: to recruit from schools: to raise the potential of the currently hired researchers: to transpose researchers within the company: to hire specialist from outside.

We analyzed the relationship between the 216 situations and the four alternatives, then we interviewed several managers to evaluate how our model could explain the managers decisions.

3.1.3 The Motivation of the Study

Our personnel department has, so far, been operated by non-engineers. It was difficult for them to understand details of the versatile technologies employed in complex electric and electronic research.

Therefore, a committee on "skills development and interchange of engineers" was established, with the authors of this paper as members. The problem was to utilize the limited number of human resources in the R & D organization most efficiently, and we are intereted in formulating logistics of manpower for different situations required.

3.1.4 The Scope of the Study

The paper describes a long range R & D skills development model to be used in the laboratories of integrated electric and electronic research. Discussions included in this report are: the gap analysis procedure between the technological requirements of future R & D projects and the present capability of R & D personnel; an evaluation method of alternative actions to fill the gap; a comparison between the results obtained from the above method and from conventional manager's decisions, using 25 past cases.

3.2 The Background of the Study

3.2.1 Characteristics of the White-collar Market in Japan

In Japan, the white-collar worker market consists of young people graduating from school. Those white-collar workers normally join business firms as reserve men for future use as specialists. They are trained by the company to become specialists because education in universities is not so practical. They work for the same company for their life-time because of social custom.

Even though specialists in some specific field are no longer necessary, they are not fired but are transferred to other sections or retrained. They give their loyalty to the company they work for, rather than to their specialities or profession. It is difficult, if not impossible, to employ specialists from other companies.

3.2.2 Change in Economic Environment

Change in the economic environment necessitated the establishment of a new R & D strategy. Abundant manpower could have been put into R & D activity under the past high growth rate of economy, but full utilization of manpower has become of prime importance in a low growth rate economy. The electric and electronic industry is one of the most progressive industries in Japan due to technological innovations. With the progress of technological innovations some obsolete researchers may appear. They may be frustrated and may become latently jobless and may be slow in their research. Therefore, long range R & D plans must be linked with a skills development plan for R & D personnel.

3.3 Conceptual Framework

The conceptual framework of our long range R & D skills development model which is being considered for implementation in the near future is shown in Figure 1. It shows our intention to link R & D plans and skills development plans.

External factors affecting long range R & D plans of a company would seem to be found in the following four environments.

(1) social environment,
(2) political environment,
(3) economical environment, and
(4) technological environment.

From analysis of the above, the R & D planning department forecasts changes in market needs and changes in technological seeds in a long range perspective. The range of time used in the forecasting may differ from subject area to subject area depending on the characteristics of the related resarch field. For example, two years in semi-conductors and electronics is appropriate, five years in machinery, and ten years in heavy electrical machinery.

R & D activity in a business firm cannot be pursued from the point of view of pure academic interest. R & D planners of the firm must estimate opportunity gain and loss by analysing the forecasted needs and seeds.

Further, business results of the firm have a great influence on R & D plans. Usually, when a business condition declines, the R & D budget may most likely be tight. The reason is that the R & D budget is not determined by what is needed to carry out its plans, but is determined by a fixed percentage of the revenue.

In the hierarchy of R & D programs the lowest organizational level, which is called the R & D theme, corresponds to what is called the R & D unit which takes the form of either a research group or a project team. Each R & D theme which is to be assigned to the R & D unit is given a priority rating in terms of its estimated degree of contribution to the business. Resources are allocated according to the priority.

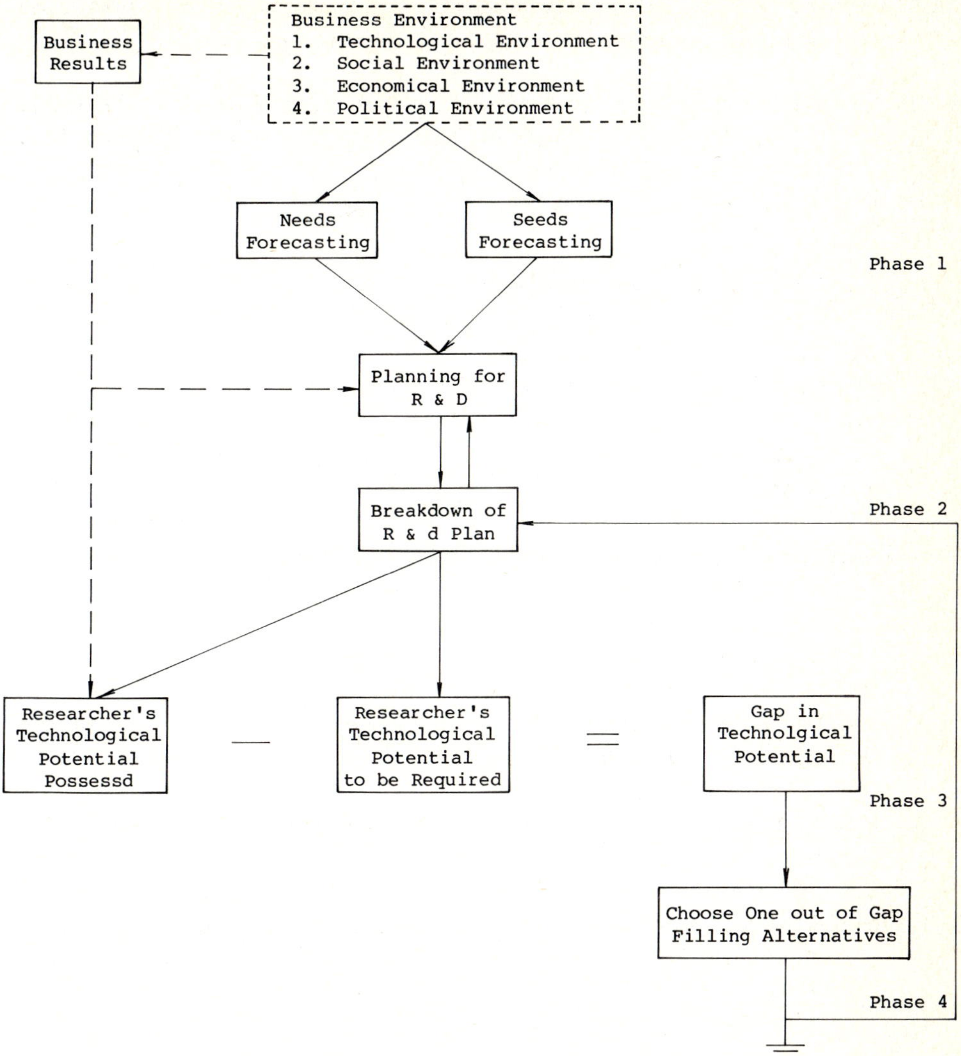

Fig. 1 Conceptual Framework for R & D Skills Development

In order to analyse the present technological potential of an R & D unit, the R & D theme is decomposed into technological specialities. The technological potential of each reseacher is also evaluated by assessing the quality of his technological speciality. The term "technological potential" is defined as "the extent of knowledge and experience of the researcher and his ability to solve problems by the use of that knowledge and experience."

Similarly, future technological potential analysis is made to predict technological specialities required to solve R & D themes. Thus we can identify the technological gap between the present and the future. According to the nature of the gap, measures to fill it are considered. Then the feasibility of the measures is checked. If the action selected is impossible, the R & D theme is to be altered, which will in turn result in a revision of the R & D plan.

3.4 Detailed Procedure of R&D Skills Development

3.4.1 Gap Analysis (phase 3 in Figure 1)

(step 1) Measurement of Technological Potential Possessed by R & D Unit.

(step 1-1) Breakdown of R & D Theme Given to R & D Unit An R & D theme is broken down into subcategories (Ti's) with which one can intelligently handle the skills mix. There are, however, no specific rules to follow in terms of how fine one should break it down. The general priciple is not too small but not too large. For example, if an R & D theme is to develop "Electron Linear Accelerator", it will be broken down as follows:

High Frequency Electronics,
Vacuum Engineering,
Electron Beam Optics, etc.

(step 1-2) Evaluation of Individual Researcher. Evaluation of the potential of each researcher in terms of Ti's is made using the following criteria:

Table 1 Evaluation criteria

Class	Value
Outstanding	1.0
Satisfactory	0.7
Below Satisfactory	0.4
Unsatisfactory	0

The result is presented as follows:

Table 2. Evaluation of individual researcher

Researcher (M_j)

	M_1	M_2	M_m
T_1	a_{11}	a_{12}	a_{1m}
T_2	a_{21}	a_{22}	a_{2m}
⋮	⋮	⋮	⋮
T_n	a_{n1}	a_{n2}	a_{nm}

a_{ij} = individual potential

(step 1-3) Evaluation of R & D Unit

The potential level of an R & D unit for a subcategory of a technology (b_i) is determined by the maximum value of the group. $b_i = \max^i(a_{i1}, a_{i2}, \dots, a_{im})$

(step 2) Assessment of Technological Requirement

To the subdivision of an R & D theme in terms of technological field of knowledge (T_i), rating value of the level is attached by evaluating the degree of difficulty (C_i) required to achieve the result.

(step 3) Computation of the Gap

The gap between the technological requirement and the potential capability of the unit (d_i) is computed by the field of technology as follows.

$$\begin{bmatrix} d_1 \\ \vdots \\ d_n \end{bmatrix} = \begin{bmatrix} c_1 \\ \vdots \\ c_n \end{bmatrix} - \begin{bmatrix} b_1 \\ \vdots \\ b_n \end{bmatrix} \quad \ldots\ldots \quad \text{Equation 1}$$

Although the technological gap ($d_1, \ldots, d_n$) was defined as a vector, in the following discussions we will use the maximum value of the vector to represent the gap of an R & D unit between the requirement and the potential.

4.2 Selection of Alternatives to fill the Gap

4.2.1 Alternatives to fill the Gap

In order to take remedial action for filling the gap which was made clear in the preceding section, there are four alternatives.

Table 3. Alternatives to fill the gap

alternatives		code
Retraining of personnel within the unit		A
Addition of personnel	hire from school	B
	obtain specialist within company	C
	recruit specialist from outside	D

4.2.2 Assessment of the Priority of Situations

The following six factors which are related to defining the situation are to be considered before the forementioned alternative actions are taken.

1. Degree of sophistication of the technology concerned
2. Urgency
3. Relative technological level of the company (as compared with the competitors)

4. Existence of specialists within the unit
5. Existence of specialists within the company
6. Allowance of manpower resources

There are various conditions that each of the above can take. These conditions are listed in Table 4.

Table 4. Possible Conditions

Factor	Condition	Symbol
1	Big Medium Small	T-1 T-2 T-3
2	Long Medium Short	E-1 E-2 E-3
3	Lower Same Higher	L-1 L-2 L-3
4	Exist Not Exist	S-1 S-2
5	Exist Not Exist	S-3 S-4
6	Abundant Tight	H-1 H-2

A situation is defined by the combination of conditions which are selected from the factors. Therefore, there are 216 (3x3 x3x2x2x2) possible situations in the above conditions.

4.2.3 Selection of Alternatives

Instead of considering the possible 216 combinations, each factor was separately treated. As will be seen in Table 5, an evaluation of the alternative actions for each condition was made. In the column of Expected Effectiveness, "0" means impossibility and "3" means the most desirable. Suppose there is a situaiton which consists of the following conditions: T-1, E-1, L-1, S-1, S-4, H-2. Then, the amalgamated value for each alternative action will be obtained by multiplying the corresponding values given in Table 5.

Table 5. Evaluation of alternative actions

Alternative Action	T-1	E-1	L-1	S-1	S-4	H-4	Total Expected Effectiveness
A		1	1	1	1	1	1
B		1	3	1	1	3	9
C		1	1	1	0	0	0
D		1	3	0	1	3	0

Alternative action B is shown to be the best.

4.3 Evaluation of the Method

4.3.1 Design of Survey

In order to examine the validity of this method, a survey was conducted by interviewing five managers of the Central Laboratory. All together, 25 cases were analyzed. For each case, one of the five managers who were concerned with it was asked to give answers to the following three questions.

1. What action did you take for each condition ?
2. Is the result satisfactory when you acted as suggested in Table 5?
3. Is the result unsatisfactory when you did not act as suggested in Table 5?

Table 6. Matrix of Conditions and Effectivieness

Symbol	CONDITION	Expected effectiveness of each of the four alternatives			
		A	B	C	D
E-1,T-1	long time allowance and big technological potential gap	1	1	1	1
E-1,Td-2	medium	1	1	1	0
E-1,T-3	small	1	1	1	0
E-2,T-1	medium big	2	1	2	3
E-2,T-2	medium	3	2	3	1
E-2,T-3	small	3	2	3	0
E-3,T-1	short big	1	1	2	3
E-3,T-2	medium	1	1	2	2
E-3,T-3	small	3	2	2	0
L-1	our company has lower technological level than other companies	1	3	1	3
L-2	same as	3	3	3	1
L-3	higher than	3	2	3	0
S-1	technolgoical specialist who can train other in R&D exists	1	1	1	0
S-2	does not exist	1	1	3	3
S-3	technogical specialist exists in the company	1	1	3	0
S-4	does not exist	1	1	0	1
H-1	human resources are aoundant in the company	3	0	3	0
H-2	tight	1	3	0	3

1. Evaluate the situation in terms of the six factors and the respective conditions given in Table 4.
2. What alternative was actually selected.
3. Evaluate the result in terms of the following four criteria.

Table 7. Result Evaluation

Evaluation	Code
Excellent	R4
Good	R3
Fair	R2
Poor	R1

4.3.2 Results of the Survey

Results of the survey are summarized in Table 8.

4.3.3 Analysis of the Result

Along with data collected from the managers,alternatives to be selected through the use of the method suggested here, are shown in Table 8. The difference between alternatives selected by the method and alternatives selected by managers were evaluated as follows.

Table 8. Difference between Alternatives

Preference ordering suggested by method	Action taken by managers				Evaluation of difference
	1	2	3	4	
1	1	0	0	0	Match
2	0	1	0	0	Slightly Unmatch
3	0	0	1	0	Unmatch
4	0	0	0	1	Totally Unmatch

This means that if the manager selects the alternative action whose preference ordering is 1, then the evaluation is

"Match"; if ordering is 2, then "Slightly Unmatch"; if ordering is 3, then "Unmatch"; if ordering is 4, then "Totally Unmatch".

Table 9 **SUMMARY OF DATA COLLECTED**

Case No	Gap Situation															Preference Ordering by This method				Alternative Selected by Manager				Evaluation of Result			
	1			2			3			4		5		6													
	T-1	T-2	T-3	E-1	E-2	E-3	L-1	L-2	L-3	S-1	S-2	S-3	S-4	H-1	H-2	A	B	C	D	A	B	C	D	R4	R3	R2	R1
1		*				*	*			*			*		*	2	9	0	0		*				*		
2	*				*		*			*			*	*		12	0	0	0	*					*		
3	*					*	*			*		*		*		3	0	18	0	*					*		
4	*					*	*				*	*			*	1	98	0	0			*				*	
5	*					*	*			*			*	*		6	0	0	0	*				*			
6	*			*				*		*		*		*		9	0	27	0			*		*			
7			*			*			*		*		*	*		54	0	0	0		*			*			
8	*					*	*				*	*			*	1	9	0	0			*				*	
9	*					*	*				*		*		*	2	9	0	81	*							*
10			*			*		*		*			*	*		27	0	0	0			*		*			
11	*					*	*				*	*			*	1	9	0	0			*					*
12	*				*			*			*		*		*	6	9	0	27	*				*			
13			*			*			*		*		*	*		54	0	0	0	*				*			
14		*			*		*				*		*		*	3	18	0	27		*				*		
15	*				*			*			*		*		*	6	9	0	27	*						*	
16	*					*	*				*		*	*		6	0	0	0	*						*	
17	*				*			*			*		*		*	6	9	0	27			*				*	
18	*				*			*			*		*		*	6	9	0	27			*					*
19	*					*	*				*		*		*	2	9	0	81		*					*	
20	*				*			*			*		*		*	6	9	0	27		*			*			
21		*				*		*		*			*		*	3	9	0	6	*					*		
22		*		*			*			*		*			*	1	9	0	3		*			*			
23	*				*			*			*		*		*	6	9	0	27			*					*
24	*			*				*		*		*		*		9	0	27	0			*			*		
25	*				*			*			*		*		*	6	9	0	0			*					*

In Table 10, results of the analysis were summerized. The calculated value of the chi-square of Table 10 is 16.63. The corresponding chi-squre distribution has 9 deg. of freedom, and the corresponding value of the upper 10 percent is 14.68. Since 14.68 < 16.63, there is statistically significant evidence.

Table 10 Result of the Analysis

	Poor	Fair	Good	Excellent
Match	0	1	2	4
Slightly Unmatch	0	1	3	3
Unmatch	2	3	1	1
Totally Unmatch	3	1	0	0

The result obtained through the use of the suggested method can explain relatively well the managers' decisions. The coefficient of rank order correlation was found to be 0.694. It seems as though managers' decisions can be improved by theuse of the suggested method.

3.5 Remarks

The forementioned method has not yet been tested in actual practice. We have already encountered some difficulties to be handled in the near future. Some of those are listed below.

1) How to motivate R & D unit managers to release their subordinates for the sake of the development of their skillseven if they are greatly contributing to his group.
2) How to compromise between a tenure dependent promotion system and a skills development system.
3) How to determine the optimum number of transitions and intervals in order to maximize a researcher's total output.

4. CONFLICTS BETWEEN MERIT SYSTEM AND SENIORITY SYSTEM IN LIFE LONG EMPLOYMENT SOCIETY

4.1 Background

The principal feature of Japanese firms is lifelong employment with seniority system, where conflicts exist when merit system is introduced, especially among research and development personnel. Scenarios of conflicts are shown for 8 research and development managers from innovative vs. adaptive technology fields, individual vs. team work styles and young vs. middle age managers.

The characteristics of the employment system in Japanese firms are lifetime employment, seniority system and are incorporated labour union. 1),2) In research and development laboratories merit system is going to be adopted because researchers' ability is more important than experience.

In the introduction period of the merit system like now, there exists conflict between the merit system and the seniority system even in laboratories, which will be dealt with in this paper.

4.2 Seniority system in Japan and Japanese Firms

In Japan social classes were abolished after World War II and the equality of people was well established by the destruction of the aristocracy and by the appearance of democracy. But the seniority system has long been a part of the mentality and public order of Japanese people, affected mainly by Confucianism.

Therefore, deviation from the seniority system in a business society leads to conflict in social life, such as in the seating order.

Spearman's rank (Chairman of the Board, President, Executive Vice-president, Senior Vice-President, Managing Director, Director) correlation coefficient was calculated between executives' rank and age for the top 53 companies in Japan, and there is a strong correlation between them, as is shown in Fig. 1. The correlation matrix was made of the coefficient, years since founding of the company, number of its employees, average age of the employees, average age of its executives and the number of its executives.

But no correlation coefficients in the matrix in Table 1 were significant, which means Spearman's rank correlation coefficients between executives' rank and age was found significant, independently of other coefficients.

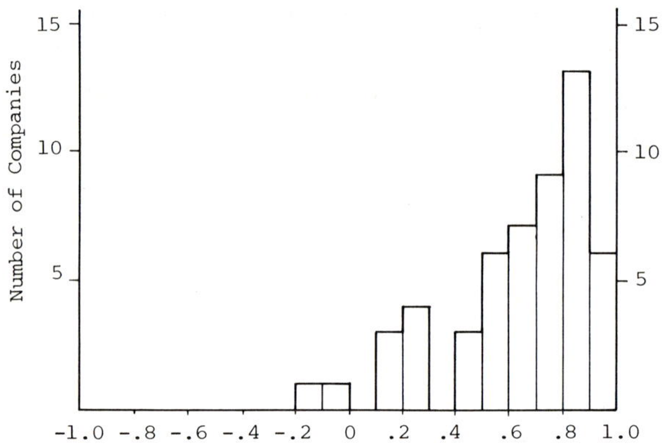

Fig. 1 Spearman's rank correlation coefficient

Table 1. Correlation Matrix

factor \ factor	1.	2.	3.	4.	5.	6.
1. Years since founding of the company						
2. Number of its employees	-.015					
3. Average age of its employees	-.026	.055				
4. Spearman's rank correlation coefficient (executives' rank and age)	.110	.220	.037			
5. Average age of its executives	.020	.418	.454	.006		
6. Number of its executives	-.032	.533	.245	.347	.341	

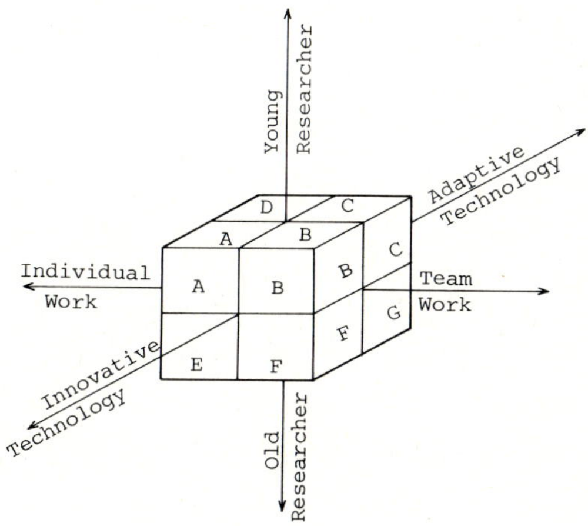

Fig. 2

4.3 Introduction of Merit System

The merit system is going to be adopted for the following reasons.

1. The seniority promotion system has become hard to maintain due to increased university graduates.
2. The seniority wage system can not be maintained due to increased starting salary and delayed retiring age owing to the lengthened life expectancy of the Japanese.
3. Promotion of young engineers is necessary in fast developing technology fields such as semiconductors and computers.

4.4 Conflicts between Merit System and Seniority System

When competent young researchers are promoted as research managers, some of them can not manage their research group well due to resistance from elder subordinates. Besides, the young managers are left out by their contemporaries, who are often a good information source in a Japanese firm, due to jealousy. Then some young managers become neurotic. Some of them quit their company and find jobs in a university. It is rather difficult to changecompanies, because the Japanese labour market is not wide open for white collar workers.

The conflicts differ by the technology fields whether innovative or adaptive, the work style whether individual work or team work, and the age of managers whether young or middle aged. By the combination of these three axis, eight managers can be described as the representatives of a laboratory. Their views for the seniority system, merit system and lifelong employment system, their feeling for promotion and their attitude for conflicts are presented by scenario writing.

4.5 Scenarios of Conflicts between Seniority System and Merit System

1. Researcher (not true name,) age, (age when promoted to manager)

2. Educational background
3. Present job
4. Character
5. View for seniority system (hereafter S.S.)
6. View for merit system (hereafter M.S.)
7. Speed of promotion compared with his contemporaries
8. Feeling if promoted faster than his contemporaries
9. Feeling if promoted slower than his contemporaries
10. View for younger boss
11. View for younger subordinate who is more competent than him
12. View for older subordinate who is disobedient to him
13. View for lifelong employment (hereafter L.E.)

1. Dr. Ando. 34, (33)
2. Mathematics
3. Computer
4. He has sharp intuition, flush or wit and confidence
5. I feel confident and full of energy in M. S.
6. I don't wish to work under an incompetent boss in S. S.
7. Fast, for my ability was superior to my contemporaries
8. It was natural and I don't feel happy.
9. I would go to my boss and protest.
10. I will work under the boss if he is competent.
11. I will compete with him fair and square.
12. I will transfer him to another section.
13. I don't want to work after the peak of my ability, and I will work in a university.

1. Dr. Baba. 38, (34)
2. Electronics
3. Semiconductor
4. He devotes himself to research and investigates very deeply.
5. S. S. is completely nonsense in a rapidly expanding field like semiconductor.
6. M. S. is very dynamic.
7. Fast, for I am competent and our field is rapidly expanding.
8. I don't feel excited but calm.
9. I would become very mad.

10. It is easy to talk frankly to a young boss.
11. I will positively help competent subordinates.
12. In the company, position talks but not age.
13. I want to work for the company in the sales department even after my technology becomes obsolete.

1. Mr. Chiba. 39, (38)
2. Mechanical engineering
3. Satellite
4. He is adaptable to a new environment
5. S. S. can adapt features of M. S. because importance of positions differs even though rank is the same.
6. M. S. with survival match is better than tournament only.
7. Slow, for my rivals were too strong.
8. I would feel proud.
9. I became a grumbler.
10. I will negotiate with my boss for the territory and authority of my job, and I wish to work independent of him, usually.
11. I will give him a separate job and make him independent of me.
12. I will censure and torment him till he obeys me or quits the company.
13. I prefer L. E., becuase welfare and education systems in the company are appreciable.

1. Mr. Doi. 39, (35)
2. Electronics
3. Communication Equipment
4. He accepts others' opinion flexibly.
5. Where there is competition, there are better results, therefore S. S. is obsolete.
6. M. S. is a very rational system.
7. Average, for our field expanded very rapidly, but managers are scouted from out of the company.
8. The principle that the stronger prey upon the weaker rules this world.
9. I will surpass the promoted at next chance.
10. He can manage better than an old manager for he knows up-to-date technology.
11. I will help to promote him.

12. I don't expect him to work in the front for his technology is oldfashioned.
13. I want to establish my own company.

1. Dr. Etoh. 43, (39)
2. Chemistry
3. Material
4. He is patient, a man of quiet taste and works hard at his jobs.
5. S. S. saves energy for competition and we can research peacefully.
6. M. S. lacks a judging rule that everyone can consentto, therefore, I prefer S. S.
7. Slow, for I was not in a place in the sun.
8. I would feel not so bad.
9. I felt jealous.
10. I will feel strange and unharmonious.
11. I will try to keep good terms with hims.
12. I will save his face
13. I want to work for the company as long as possible.

1. Mr. Fukuda. 45, (36)
2. Physics
3. Nuclear Engineering
4. His way of thinking is logical and systematic.
5. S. S. is suited for long range research.
6. M. S. makes researchers feel uneasy.
7. Average and I am satisfied and feel no stress.
8. Troubles will increase among contemporaries.
9. I had heavy feet to attend meetings where the fast-promoted would come.
10. If he is more competent and has better achievement than I, I am not discontented.
11. I will produce an atmosphere where no age differences are felt, and we can cooperate in peace and harmony.
12. I will make democratic rules fitted for him to work easier even under a young boss.
13. I will work for the company unless there is demotion or salary reduction.

1. Dr. Gotoh. 47, (38)

2. Electrical Engineering
3. Heavy electric machine
4. He looks at the main issue of things.
5. S. S. is superior to M. S. because education in Japan is not a genius breeding education. Therefore, M. S. is dangerous where abilities differ little.
6. In M. S. one-man play, sometimes stand play, is shown rather than team play.
7. Average, for our field is stable.
8. I would feel uncomfortable.
9. I would feel disappointed.
10. I will be on hard terms with him.
11. I will transfer him to other section.
12. I will keep him at a respectful distance.
13. Even when our company is in hardship, we should endure equally and keep L. E.

1. Mr. Honda. 46, (35)
2. Mechanical engineering
3. Mechanical structure analysis
4. He has good human relations and large circle of friends.
5. Requirement for the boss is not always limited to first class technology, but management ability, experience and personality can lead his subordinates.
6. As a whole M. S. is inefficient because only a few promoted are motivated, but many unpromoted feel depressed and less cooperative.
7. Fast, for my boss was promoted.
8. I felt critical eyes would be sharper with the rise of my status.
9. I would feel ashamed and keep it secret to my wife.
10. I can not bear to take a lower seat than a younger boss, therefore, I will apply for transfer to other section.
11. I would avoid head-on collision with him.
12. I will respect his experience and try to make him my advisor.
13. In L. E. company, boss can transfer his whole technology to his subordinates, without fear of competition.

4.6 Consideration

Although the seniority system is a part of the mentality of Japanese, the merits system will become popular among researchers whose ability is more important than age.

Therefore, during the transient period like now there are two countermeasures as follows.

1) Line and staff should be separated, and competent young candidates for managers should be promoted.

2) Virtual merit system should be adopted by appointing competent young candidates for managers of projects.

Reference

1) J. C. Abegglen, The Japanese Factory - Aspects of its Social Organization, Free Press. 1958.

2) P. F. Drucker, What we can learn from Japanese management P110-122, Harvard Business Review; March-April 1971.

5. PRE-EVALUATION OF RESEARCH AND DEVELOPMENT PROJECTS -- COMPARISON OF ECONOMY CALCULATION AND A PROPOSAL OF A NEW METHOD --

5.1 Purpose

Qualitative methods have been used to select research and development (hereafter R & D) projects in our laboratory.

In the low economic growth age after the oil shock, requests for quantitative evaluation of R & D activities became strong to find effective R & D investments.

This paper reports an approach for quantitative evaluation, especially economy calculation, its application to a model and proposal of a new method.

5.2 Evaluation of R & D projects

5.2.1 In evaluating R & D projects, there are eight aspects as is shown in Table 1.

Table 1. The aspects of R & D project evaluation.

View Point / Point of Evaluation	Economical Evaluation	Technological Evaluation	Social Evaluation	Political Evaluation	Environmental Evalua tion
Pre-Evaluation	A1	A2	A3	A4	A5
Intermediate Evaluation	B1	B2	B3	B4	B5
Post Evaluation	C1	C2	C3	C4	C5

Both qualitative and quantitative evaluation are involved in the evaluation of A1 ∿ A5, B1 ∿ B5 and C1 ∿ C5. Although many companies intended to change from qualitative to quantitative evaluation, improvement of data input is necessary.

B.V. Dean developed the scoring method by improving quantitative evaluation.

Economical evaluation of R & D activities, i.e. cost effectiveness analysis, is popular in pre, intermediate and post evaluation.

Technological evaluation is done not only in intermediate and post evaluation but also in pre-evaluation; one typical method is technology assessment.

5.2.2 General Scope of Evaluation Method: Many evaluation methods have been proposed for R & D projects. A.H. Rubensteis developed a method of classification of the methods as follows.

1) Decisive evaluation: Evaluation items and criteria are set, then ratings by intuitive decision are made. Adoption or rejection of R & D projects and priority of those adopted is decided.

2) Economical evaluation: R & D outputs are evaluated from the view point of economy by the ratio of cost and benefit.

3) Evaluation by operation research: Representing every event in R & D activity by mathematical model and changing elements dominating the event multidimentionally and dynamically, the future of the activity is forecasted.

5.3 Economical Evaluation

5.3.1 Principle and feature of economical evaluation.

In economical evaluation profitability of R & D is quantitatively measured as the economy index which is a parameter of R & D evaluation.

Two types of economy indexes are as follows.

Economy index (I) = Output/Input
Economy index (II) = Output-input

The former is a relative measure and the latter is difference measure.

The R & D cost or expense which is input is defined by "The Survey of R & D in Science and Technology" by the Japanese Government as follows.

1) Research cost = human cost + material cost + depreciation of tangible fixed asset + document cost + other cost

2) Research expense = human cost + material cost + initial cost of tangible fixed asset + document cost + other cost

It is necessary to define the extent of R & D expenditure although either cost or expense is adopted as input. The calculation method of R & D expenditure must be standardized and manualized.

In similar way, R & D output must be calculated in monetary measure in the economical evaluation method. The output is usually measured by profit of sales and sometimes by marginal profit or incremental profit.

Imaginary losses to be expected if R & D for some projects is not performed were considered to be opportunity profit, and they are regarded as R & D output.

Besides, there are some concepts to see equivalent royalty and equivalent service as output; the former is royalty to give in case a certain technology is bought from another company, and the latter is estimated cost if certain technological service is purchased from outside the company.

Economical evaluation methods can be classified into two groups i.e. index formula method and economy calculation method. The former is to make a formula to get evaluation indexead by simply combining evaluation factors and to evaluate by using the indexes calculated by the formula as parameters. The latter, on the contrary, is to look at R & D expenditure as an investment and to evaluate economy of R & D positively by introducing decisive concepts such as present value and profit on investment.

5.3.2 Index formula method: Evaluation by index formula is not sufficient theoretically, but it is practical because of relatively simple usage.

There are Olsen method, Pacifico method, Teal method, return on research index method, O'meara method, Hertz method, Sobelman method etc.[1)]

5.3.3 Economy calculation method: Various methods are presented to select R & D investment by economy claculation. They are basically divided into three categories as follows.

1) comparison of amount
2) comparison of ratio
3) comparison of period

The first one is to use net profit considering capital cost and is named "present value method."

The second method is to apply the concept of effectiveness for long range investment and is named "return on investment method."

The third method is called "period of return method."

These three methods are related to each other. If we consider the present value method as standard, the return on investment method is to obtain a reduction rate to put the net present value at zero in the present value method.

If the reduction rate of the present value method is made equal to the interest rate, the period of return method becomes equal to the present value method. But the period of return is difficult to use for evaluation because R & D investment is usually continuous, the R & D period is versatile and the maximum permissible period of return can not be fixed rationally as evaluation criteria. Therefore, the ratio of the present values of profit and the cost of R & D is employed as evaluation parameters by the name of "project value index" or "project number."

Here we call the parameter "project index."

(1) Present value method

Similar to the present value, there exist a final value and an annual value. Among present value P, final value S, and annual value M there is a relation that $S_A < S_B$ and $M_A < M_B$ if $P_A < P_B$ and the interest rate and period are the same. It is convenient to know the size of the project if R & D expenditure is converted into the present value and we use the present value as representative.

In the present value method, future net cash flow for R & D expenditure and profit is estimated, net present value is calculated and projects are selected. It is necessary for a project to be selected that at least its present value is positive (NPV>0).

The present value defined by J. B. Quinn or B. V. Dean and S. S. Sengupta is shown as follows.

$$V = \sum_{i=0}^{n} [Ci \ (1 + r)^{-i}] \ . \ . \ . \ . \ . \ . \ . \ . \ . \ . \ . \ . \ . \quad \text{Eg. (1)}$$

Ci : Net cash flow in i-th year

n : all the years when profit is expected

r : reduction rate (interest rate to be expected)

It is better for the reduction rate to use the profit rate to be expected for the project instead of a fixed rate. Projects will be carried out in order that their project present value is thus calculated.

(2) Investment rate of return method

Using the investment rate of the return method we can get the profit rate "r" (internal rate of return: IRR) which satisfies the following formula and we can evaluate R & D projects.

$$Pc \cdot Pt \sum_{k=1}^{n} [I_k (1+r)^{-k}] = \sum_{h=e}^{-m} [Oh \ (1 + r)^{-h}] \ . \ . \ . \quad \text{Eg. (2)}$$

where Pc, Pt are commercial and technological probability of success, respectively

I_k : profit of project in k-th year

Oh : investment for project in h-th year

n : total years when profit is expected

ℓ,m : starting and finishing years, respectively

The standard year for calculating IRR can be chosen arbitrarily. The Eg. (2) uses the year when the project is industrialized and makes the profit as standard as shown in Fig. 1.

Profit is obtained at the end of the year and investment is made at the beginning of the year.

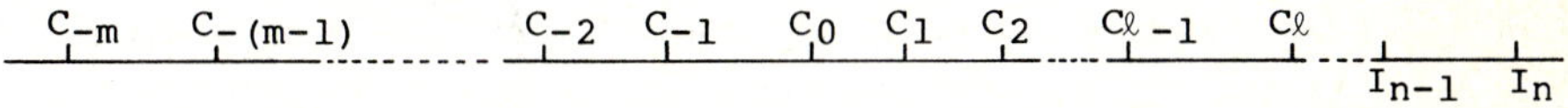

Standard year

Fig. 1 Standard year for Calculating IRR

(3) Rate of return for the investment method: Projects are evaluated in this method by using the ratio of project profit discounted to net value and R & D expenditure as the evaluation parameter. This method uses relative measure of economy index (I), whereas the net present value method uses difference measure of economy index (II) as mentioned in section 3-1.

When R & D investment is very small compared to the total investment for the project, the evaluation by the investment profit rate method is meaningless as an evaluation of R & D because of lower interest than profit than interest but the project index method is meaningful. The typical example of the project index method is Disman's project number method.

Disman thought that the maximum expenditure justified as net present value of project profit from the view point that the maximum R & D expenditure should be no more than expected profit.

Therefore, the maximum expenditure justified (hereafter MEJ) is theoretically as follows.

$$MEJ = \sum_{k=1}^{n} [I_k(1+r)^{-k}]_k \quad \text{. . . . . Eg. (3)}$$

where I_k : Project profit in k-th year

n : total years expecting profit

r : reduction rate (expected profit rate)

Considering probability for success, the maximum expenditure justified for improving the manufacturing process, MEJ, is as follows.

$$MEJ_I = P_t \sum_{k=1}^{n} [I_k(1+r)^{-k}] \quad \ldots\ldots\ldots \text{Eg. (4)}$$

and the maximum expenditure justified for developing new products and new technologies, MEJ_p, is as follows.

$$MEJ_p = P_c \sum_{k=1}^{n} [I_k(1+r)^{-k}] \quad \ldots\ldots\ldots \text{Eg. (5)}$$

where I_k : estimated cost reduction due to process improvement.

Pt. Pc : probability for technological and commercial success, respectively.

On the other hand, the total cost for R & D, TC, is as follows if the fruit of R & D is industrialized immediately after R & D.

$$TC = \sum_{k=1}^{-n} [O_h(1+r)^{-n}] \quad \ldots\ldots\ldots\ldots\ldots \text{Eg. (6)}$$

O_h : R & D expenditure for the h-th year

m : the total years expending for R & D

Therefore, the project number which is the ratio of MEJ and TC, is as follows.

$$PN = Pc \cdot Pt \sum_{k=1}^{n} [I_k(1+r)^{-k}] / \sum_{n=-1}^{-m} [O_h(1+r)^{-h}] \quad .. \text{Eg.(7)}$$

where, if PN=1, we can expect an investment profit rate equal to reduction rate r,

If PN 1, we can expect an investment project rate higher than reduction rate r,

If PN 1, we can expect investment project rate lower than reduction rate r.

The priority of projects can be decided by the difference of PN.

5.4 Study of Cases and Models

Present cases are for R & D projects for cost reduction of facility equipment.

In general, when we calculate the economy of projects there are several problems such as

1) the difficulty in evaluating the fruits of R & D
2) uncertainty in probability for success of projects.

In this case the project is for cost reduction and its fruit is clearly measurable.

Need forecast of facility equipment can be done relatively accurate compared with that of home appliances. (Forecast was done by sales department in this case.) When the cost reduction project was completed, the manufacturing line was changed and we thought (cost reduction for a product) x (number of products sold) = profit

Different from the new product development project, the cost reduction project can generate some extent of result irrespective of success or failure,which is a merit of the cost reduction project.

In this case, therefore, the success probability for R & D was assumed to be unity in calculating economy. It was also assumed that the profit harvesting period for the R & D is five years.

In order to test the hypothesis for the case, nine models for R & D investing were employed, and the relation between the annual investment pattern and the calculated value of economy were studied in general.

5.4.1 Original series of the cases

Fig. 2 shows the cash flow of projects A through 1.

Investments (Cn) are shown or semiannual data and profits (Vn) are shown as annual data. The Cn and Vn are shown logarithmically and the standard year is shown as zero; solid lines show investments and dotted lines show profits.

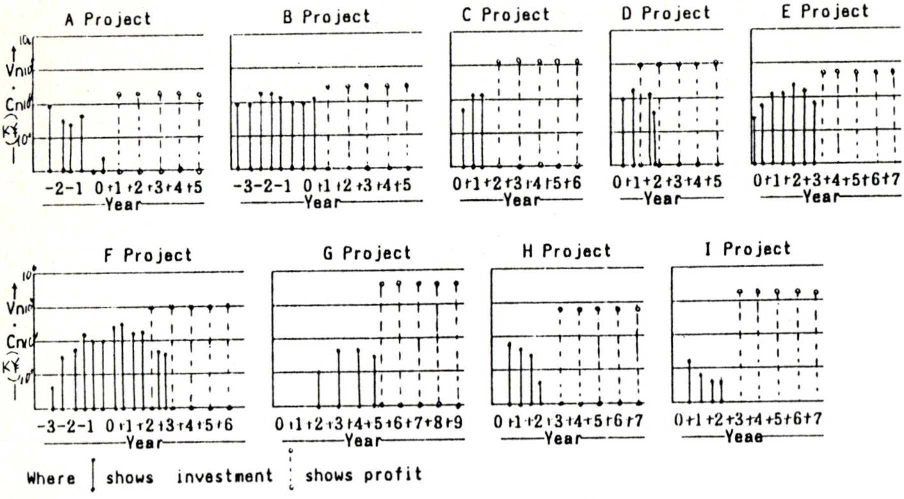

Fig. 2. Original Series of the Cases

5.4.2 Results of the calculation for the cases

Evaluation index and evaluation order by several calculation methods are shown in Table 2.

Table 2. Results of the calculation for the cases

Project	Net Present Value (NPV)	Evaluation Order	Investment Return Rate (IRR)	Evaluation Order	Project Number (PN)	Evaluation Order	Net Value for Total Project ΣVn	Evaluation Order	Net Value for Total Investment ΣCn	Evaluation Order
A	66240	8	0.54	6	3.85	6	89512	9	23272	3
B	9413	9	0.12	9	1.07	9	136162	8	126749	8
C	434592	3	2.22	3	15.95	4	463652	3	29060	4
D	244658	4	3.45	2	7.13	5	284568	4	39910	6
E	118496	7	0.44	7	2.57	7	193980	7	75484	7
F	147840	6	0.36	8	2.14	8	277874	5	130034	9
G	1357216	1	6.00	1	150.00	1	1366325	1	9109	1
H	183927	5	1.54	5	16.10	3	196095	6	12168	2
I	574734	2	1.60	4	20.3	2	604577	2	29843	5

In Table 1, the evaluation order of Cn starts from the smallest one.

Internal return rate (R) in the investment return rate (IRR) method is obtained by solving general non linear equation of the Newton repetition method using a computer or by plotting R and PN on a graph, and we can find a point where PN = 1 by trial and error.

Table 3. Correlation coefficient matrix between economy index

	NPV	IRR	PN	Σ Vn	Σ Cn
NPV		.883	.949	.995	-.513
IRR			.861	.862	-.599
PN				.948	-.450
ΣVn					-.427
ΣCn					

5.4.3 Consideration of the result of calculation for the cases

Results of calculation of the evaluation index for net present value (NPV), investment return rate (IRR), and project number method (PN) are shown in Table 2.

The evaluation orders of NPV, IRR and PN are almost coincidental, but for some projects evaluation orders are different, as shown in Table 2.

Table 3 is a correlation coefficient matrix to check the mutual coincidence of the evaluation index in Table 2. Relatively high correlation is observed for NPV. IRR and PN in Table 3, which shows similar results for evaluation and profitable projects in the total cash flow, have high evaluation by each method, while less profitable projects have low evaluation.

The correlation coefficient between NPV and V_n is very high: 0.995. V_n and NPV seem to have a close relation. But from the definition of NPV and PN.

$$NPV = \Sigma V_n \left(1 - \frac{1}{PN}\right).$$

In the present case PN is relatively large (therefore 1/PN is small) and consequently the correlation coefficient between NPV and V_n became high.

Now we consider the difference of evaluation order by each method, as shown in Table 2.

We use notation $X \leftrightarrow Y$ for reverse order of evaluation of X and Y.

Change of evaluation order in NPV and IRR : $A \leftrightarrow F$, $D \leftrightarrow I$
Change of evaluation order in NPV and PN : $A \leftrightarrow F$, C, H, D
Change of evaluation order in IRR and PN : C, D, H, I

Changes of evaluationorder were seen in the projects of A, C, D, F, H and I. Evaluation order for V_n and C_n, which are the bases of calculation in each method, is different in each project. We can, therefore, make a hypothesis that difference in evaluation order for each method will be generated when there is difference in relative evaluation order in C_n and V_n. --------- (Hypothesis 1)

We can simply classify annual trend curves of R & D investment of A,C,D,F,H and I projects which have a difference in evaluation order into three types of P, Q and R, as shown below.

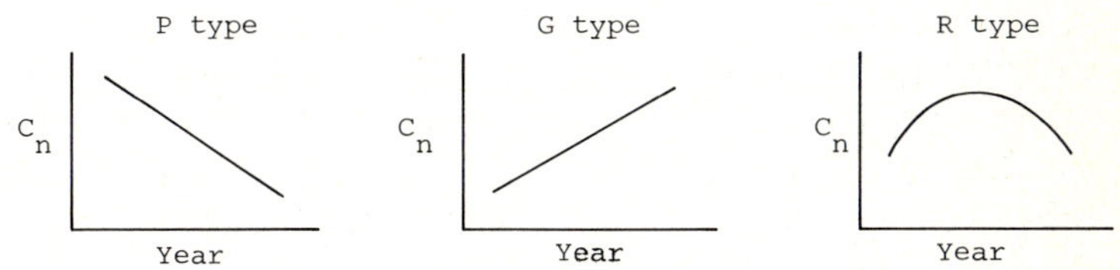

A.H. and I project C project D and F project

Fig. 2. Annual trend of R&D investment

If we review the changes of evaluation order in Table 2 and the three types using notation of ↑ to show over-evaluation compared with other methods and ↓ to show under-evaluation,

For NPV and IRR	A(NPV↓ IRR↑)	⟷	F(NPV↑ IRR↓)	... ①
	D(NPV↓ IRR↑)	⟷	I(NPV↑ IRR↓)	... ②
For NPV and PN	A(NPV↓ IRR↑)		F(NPV↑ IRR↓)	... ③
	C(NPV↑ PN↓),		D(NPV↑ PN↓),	
			H(NPV↓ PN↑)	.... ④
For IRR and PN	C(IRR↑ PN↓),		D(IRR↑ PN↓),	
	H(IRR↓ PN↑),		I(IRR↓ PN↑)	.... ⑤

From the above review of ①, ② and ③ we can get a hypothesis that evaluation order of NPV and IRR or NPV and PN may change when we compare projects of different types. ----- (Hypothesis 2)

Form the above review of ③ and ④, we can get a hypothesis that NPV is under-evaluated and PN is over-evaluated for P type project. ----- (Hypothesis 3)

From the above review of ③, ④ and ⑤ we can get a hypothesis that PN is over-evaluated for P type project -----(Hypothesis 4)

From the above review of ④ and ⑤ we can get a hypothesis that PN is under-evaluated for Q type project -----(Hypothesis 5)

Nine models of investment type were assumed in order to study the relation between annual investment types and economy indices of NPV, IRR and PN. The amounts of investment for the projects are decided intentionally, and annual tendencies of investment are forecasted comparatively accurately, but returns from the projects are only estimated and are of a relatively monotonous type. Therefore it is less meaningful to study annual tendencies of returns, and we adopted them only when V_n is apparent for return.

5.4.4 Setting of models

Models of investment type are set up for on interest rate of 10% and a standard year or the 6th year.

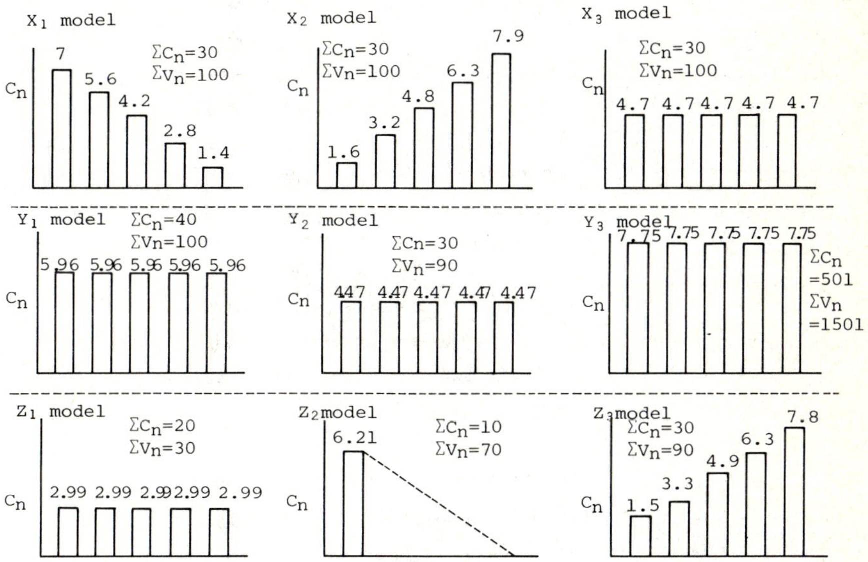

Fig. 3. Models

The relation among the three methods can not be generalized because IRR can not be obtained analytically. Examples to show the relation among the three methods are the models shown in Fig. 3. The economy indices are calculated as shown in Table 4.

Table 4.

	Net present value NPV	Investment return ratio IRR	Project number PN	Net present value of total investment Σ Cn	Net present value of total return Σ Vn
X_1	70	0.486*	3.33	30	100
X_2	70	0.675*	3.33	30	100
X_3	70	0.550*	3.33	30	100
Y_1	80*	0.507	3	40	120
Y_2	60*	0.507	3	30	90
Y_3	100*	0.507	3	50	150
Z_1	60	0.623	4*	20	80
Z_2	60	0.623	7*	10	70
Z_3	60	0.623	3*	30	90

* shows items with change

where $NPV = \Sigma Vn - \Sigma Cn$

IRR is internal rate of interest when Vn = Cn

$$PN = \frac{\Sigma Vn}{\Sigma Cn}$$

Though IRR can not be obtained analytically, NPV and PN have the following relations.

$$NPV = \Sigma Vn \ (1 - \frac{1}{PN})$$

$$NPV = \Sigma Cn$$

$$PN = \frac{\Sigma PV}{\Sigma Cn} - 1$$

$$PN = 1 - \frac{\Sigma Vn}{NPV}$$

5.4.5 Conclusion from the models and verification of hypotheses

(1) When there is no difference in NPV or PN of models X_1, X_2 and X_3, there is a relation as follows.

$IRR(X_1) < IRR(X_3) < IRR(X_2)$

(2) When there is no difference in NPV or PN of models Z_1, Z_2 and Z_3, there is a relation as follows.

$PN(Z_3) < PN(Z_1) < PN(Z_2)$

And hypotheses 3, 4 and 5 are adopted.

(3) When projects of different types are compared like X_1, X_2 and X_3 or Z_1, Z_2 and Z_3, there are cases where a difference appears in the evaluation order of IRR or PN. And hypothesis 2 is adopted.

(4) When there is a difference between Σ Cn and Σ Vn, i.e. NPV or in the ratio between them, i.e. PN, like Y_1, Y_2 and Y_3 or Z_1, Z_2 and Z_3, there is not always any difference in the other indices. And hypothesis 1 is rejected.

5.4.6 Study of uncertainty; sensitivity analysis

Uncertainty in forecast is always a problem not only in R & D investment but also in business planning.

The uncertain elements which need to be forecasted in our case are demand as the basis for return calculation and years of continuous sales. The resistivity of NPV against uncertain elements are calculated.

The decrease of NPV is calculated when uncertain elements of demands and years of sales are less than the estimation, as shown in Table 5.

When demands and years of sales were 20% less than the estimation, the decrease of NPV is affected more by the decrease of demand than by that of years of sales.

Therefore, accuracy of the forecast is more necessary for demand than years of sales. General study for this is shown below.

Equation (1) is obtained in calculating economy when we use the notation of P for net present value of total return, r for interest rate, N for years of sales and M for demand.

$$P = M \times \frac{(1 + r)^N - 1}{r(1 + r)^N} \quad \ldots\ldots\ldots\ldots\ldots \quad (1)$$

Table 5. The rate of decrease in NPV when demand and years of sales are decreased by 20%

Project	Original NPV	NPV' in 20% less sales years	$\frac{NPV'}{NPV}$	$1 - \frac{NPV'}{NPV}$ (X 100%)	NPV" in 20% less demand	$\frac{NPV''}{NPV}$	$1 - \frac{NPV''}{NPV}$ (X 100%)
A	66240	50192	0.76	24%	48338	0.73	27%
B	9413	-15226	-	-	-17819	-	-
C	434592	353750	0.81	19%	341862	0.79	21%
D	244658	194233	0.79	21%	187744	0.77	23%
E	118496	84176	0.71	29%	79700	0.67	33%
F	147840	98349	0.67	33%	92265	0.62	38%
G	1357216	1116214	0.82	18%	1083951	0.80	20%
H	183927	149319	0.81	19%	144708	0.79	21%
I	574734	467004	0.81	19%	453819	0.79	21%

As N, M and r are forecasted they have some uncertainty. The degree of influence of forecast error in M, N and r on P is related to the necessary accuracy forecast for M, N and r, and is studied hereafter.

When we totally differentiate Eq. (1), we can get Eq. (2)

$$\frac{\Delta P}{P} = \frac{\Delta M}{M} + \alpha \frac{\Delta r}{r} + \beta \frac{\Delta N}{N} \quad \ldots\ldots\ldots\ldots\ldots \quad (2)$$

Where and is expressed by eq. (3) and (4), and their value can be obtained from Fig. 3(a) and (b).

$$\alpha = \frac{N \cdot r \cdot (1 + r)^{N-1}}{(1 + r)^{N} - 1} - \frac{N \cdot r}{1+r} - 1$$

$$\beta = \frac{N \cdot (1 + r)^{N} \ \ell n(1 + r)}{(1 + r)^{N} - 1} - N \ell n(1 + r) \quad \ldots\ldots\ldots \quad (3)$$

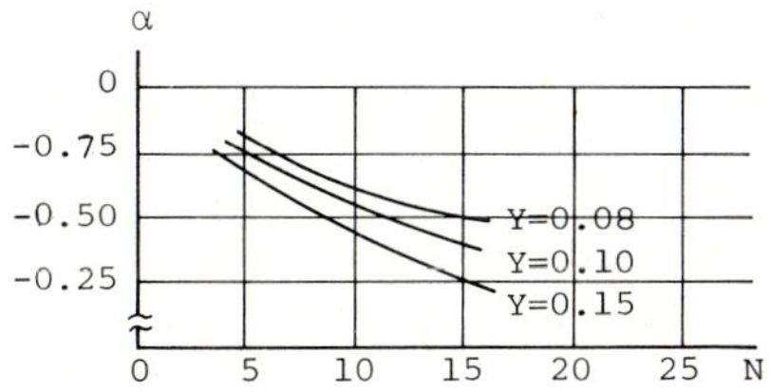

Fig. 4 (a) Characteristic of α

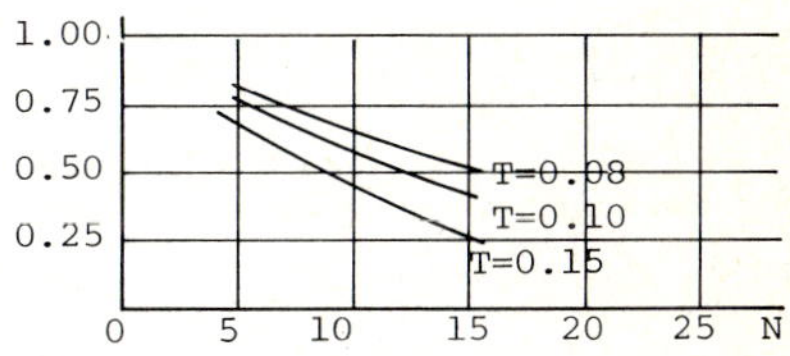

Fig. 4 (b) Characteristic of β

The ratio of $\Delta P/P$, $\Delta M/M$, $\Delta r/r$ and $\Delta N/N$ in Eq. 2 are the error ratio of net present value of total return, demand, interest rate and years of sales respectively.

The first term on the right hand side of Eq. (2) has a coefficient of unity, which implys that the error ratio in demand directly affects the error ratio of net present value of return.

The second term on the right hand side of Eq. (2) has a coefficient of α and $-1 < \alpha < 0$ for the range shown in Fig. 4(a). The absolute value of α increases with the increase of years of sales, which implys that the effect of $\Delta r/r$ on $\Delta P/P$ increases with the increase of N.

The third term on the right hand side of Eq. (2) has a coefficient of β and $0 < \beta < 1$ for the range shown in Fig. 4(b). The value of β shows a monotonous decrease with the increase of years of sales, which implys that the effect of N/N on P/P decreases with the increase of N.

From the above study we can know that the order of effect on the present value of the total return is interest rate r, demand M and the years of sales N; therefore accuracy of forecast is necessary in the order of interest rate r,demand M and years of sales N.

5.5 Conclusion

Three methods have their own features, as shown in Table 6. From the stand point of business firms, it is necessary to use them together.

The problem is, as we studied, there is a difference in evaluation order depending on investing types. How to treat the difference is a great question.

In order to answer this question, we offer a new index to integrate those three methods as follows.

$$Y = A\ (NPV)^{a} \cdot (IRR)^{b} \cdot (PN)^{c}$$

where Y : integrated index of evaluation

a, b, and c are factors for NPV, IRR and PN, respectively by this integrated index, we can pre-evaluate R & D projects considering net present value, investment return ratio and project number integratedly, which was impossible by each index alone.

Reference

(1) Editorial Committee for Research and Development "Guidebook on Research and Development" (in Japanese) Nikkagiren Pub. Co. 1973

(2) B.V. Dean "Evaluation of Research and Development Projects" (translated into Japanese by Matsui and Makiyama) Japan Management Association 1970

(3) N.R. Baker an W.H. Pound IEEE Transaction on Engineering Management Vol. EM-11 No. 4 (1964) P.126-127

(4) Marrin J. Cetron "Technology Forecast" (translated into Japanese by Terasaki and Azuma) Industrial Efficiency College 1970 P.238-247.

(5) C. Heyel (ed.) "Hand Book of Industrial Research Management" Rein-hold Pub., Inc. 1960 P.298-299

(6) B.V. Dean and S.S. Sengupta "On a Method for Determining Corporate Research and Development Budgets" Management Science, Models and Techniques, C.W. Churchman and M. Verhuest (ed.) Vol.2 Pergamon Press 1960 P.210-225

(7) S. Senju and T. Fushimi "Planning of Equipment Investment" Nikkagiren Pub Co. 1974.

(8) W. Morris "Decision System of Equipment Investment" (translated into Japanese by K. Kikuchi) Toyo Keizai Shinpo 1969

(9) M. Abe and S. Murata (ed) "Empirical Studies on Product Life Cycle" Nippon Jitsugyo Pub. Co. 1969

(10) Y. Kugimiya "Basis for Dynamic Portfolio Selection" Kyusyu Univ. Press. 1976.

6. MITSUBISHI ELECTRIC LATENT DEMAND FORECAST (MELDEF) --- A DEMAND MODEL FOR NEW PRODUCTS BASED ON CONSUMERS' ATTITUDE CHANGES ---

6.1 Purpose

Successful R & D management relies largedly on the demand forecast for a product to be developed. Without a right insight into the demand, the development activities would turn out to be a waste of resources when there was not demand enough to recover the R & D costs, or there would & be a loss of opportunities when the commercialization was too late or the production capacity was too small in comparison with the demand.

In this section, a method to forecast the demand for a new product to be developed is described briefly. The method is noted for its explicit incorporation of consumers' attitude changes over time.

6.2 Attitude changes

The car is a good example to show how important the attitude changes are. Among important attributes of the car are price, size, style, and gas mileage. The recent tendency of consumer behavior indicates that more and more emphasis is being placed on the gas mileage when consumers evaluate alternative brands since the oil embargo and the resulting price hike in gas. This point is evidenced by the fact that cars with high gas mileage are being sold better, relative to gas guzzlers, than before.

In general, consumers' attitude today will not be valid in the future. The R & D manager or marketer has to forecast the consumers' attitude associated with his or her product, especially when the development takes a long time. The product planning for a new product should be based on consumers' attitude in the future when the new product is put on the market.

6.3. MELDEF

MELDEF, developed by the Soft Science Group (SSG) in Mitsubishi; Electric Corporation, is a very simple and easy-to-use demand model. The model was developed to forecast the demand for new products, mainly consumer products, based on consumers' attitude changes over time.

At the first step, a two-level tree structure of attributes for the products competitive with or in the same category as a new product under consideration is defined. Products have many attributes and some attributes can be grouped into a higher level of attributes. For example, the economy valuie usually consists of the following three attributes:

Economy value
- initial cost
- operating cost
- product life

In MEDEF, all important attributes are grouped into six higher levels of attributes which are called the value: economy, space, time, safety, comfort, and convenience. Although these six values are common to almost all types of consumer products, some minor changes have to be made from product to product.

Secondly, the utility of each product, including a new product, in the economy value, for example, is estimated based on two kinds of questionnaire to consumers who are supposed to use or make a decision to buy the products. One is about the relative importance of the three attributes in the economy value, and the other is about consumers' judgement or evaluation of each product in terms of the attributes. The total of relative importance figures should be 10, as illustrated below. The figures for consumers' evaluation of each product should be read as follows: 5-very satisfactory, 4-satisfactory, 3-indifferent, 2-unsatisfactory, and 1-very unsatisfactory.

	Relative importance of attributes	consumers' evaluation of productsa		
		product 1	product 2	product 3
Economy value				
┌initial cost	5	4	1	5
├operating cost	4	2	5	1
└product life	1	3	4	3

In the above example, the utility of product 1, 2, and 3 in the economy value is 31(=5x4+4x2+1x3), 29, and 32, respectively. The utility of each product in other values can be estimated in the same way.

The third step is for attitude changes. Consumers' attitude at a specific time is estimated with the weight of each value being treated as a proxy of consumers' attitude. The weight is used to get the total utility of each product in the fourth step from the utility in six values calculated in the second step. And consumers' attitude changes over time is derived from the changes in the weights from time to time.

Let D_{it} = demand for product i at time t
U_{ij} = utility of product i in value j, and
W_{jt} = weight of value j at time t.

Among several possible methods for estimating W_{jt}, the correlation between D_{it} and U_{ij} is used in MELDEF because of it simplicity. When the correlation of value j with demand is greater than that of value k, it would be reasonable to judge that consumers select their favorites based on value j of each product to more extent than valve k, and therefore W_{jt} is greater than W_{kt}. In addition, when the correlation in value j is increasing over time it would be reasonable to judge that consumers select their favorites based on valve j of each product to more extent than before and therefore W_{jt} is increasing.

The correlation coefficients between D_{it} and U_{ij} should be standardized so that the total becomes 100 to get the weight of each value at time t. Also an adjustment is needed for negative coefficients. In order to avoid negative coefficients, a partial correlation is desirable if the number of products is large enough statistically.

The reasoning at this step can be summarized as follows: consumers' attitude = weights of the values (W_{jt}) correlation between the demand (D_{it}) and the utility (U_{ij}).

At the fourth step, a regression analysis is applied to the following equation, and parameters are estimated using the least square method:

$$D_{it} = f(U_{it}, t)$$
$$U_{it} = \sum_j W_{jt} \cdot U_{ij}$$

One specific formula which has been used most frequently is

$$D_{it} = a\, U_{it}^{b}\, e^{ct}$$

Where a, b, and c are parameters. And e^{ct} is a natural growth factor of demand, such as the population growth or the increase in income level.

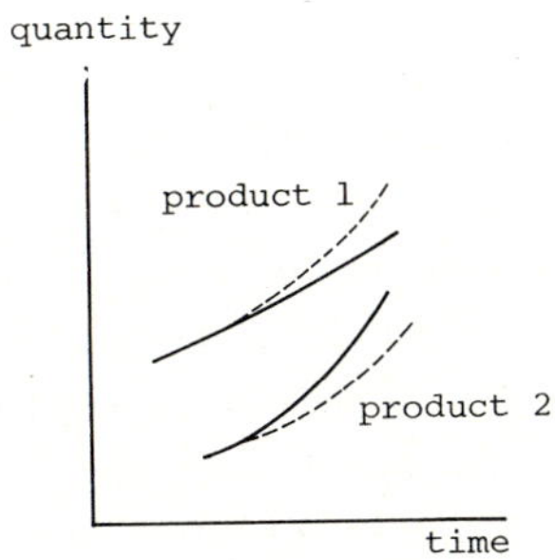

Fig. 1

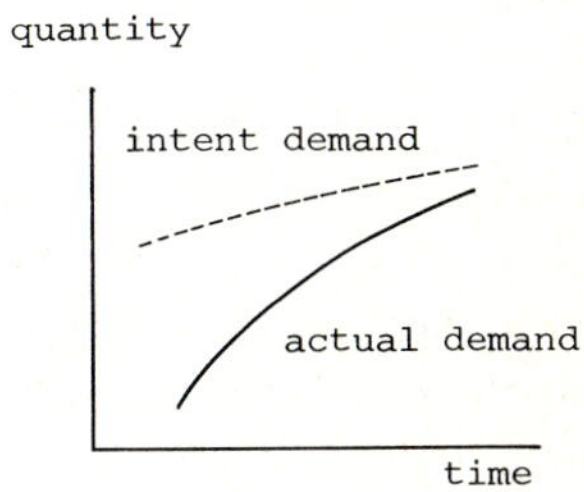

Fig. 2

An implication of the above equation is shown in Figure 1. The dashed lines show the demand for product 1 and 2 when there is no change in consumers' attitude, over time. The demand for product 1 increases at the same rate as product 2 because of the natural growth factor. Actually, however, the demand for product 2 increases at a higher rate than product 1. This is because product 2 is superior to product 1 in a value of which weight increases over time. A product meeting consumers' changing needs would increase its market share.

The next step is the extrapolation of weights of each value into the future, paying attention to forecasted socio-economic conditions associated with the product under consideration.

Once the future attitude of consumers or the weights of each value are estimated based on the past trend, the demand for each product including the new product can be forecasted using the equation above. The forecasted demand for the new product should be understood as latent demand or the maximum demand since an important factor to determine the actual demand, a supply-side factor, is not considered so far. The latent demand is not always actualized. It is efforts by suppliers to actualize the latent demand (Figure 2).

6.4 Applications in SSG

MELDEF described in this section has been applied to several consumer products under or before development since the model was developed in 1974.

One of them is MAGHEAT®, newly developed cooking equipment for household use. The study showed that in cooking equipment the importance of comfort and convenience would increase and the importance of economy would decline (Table 1). Then it was estimated that MAGHEAT® would be promising since the new product was highly evaluated in comfort and convenience although the price was rather high. The results of the study were reported to the manager in charge and soon after MAGHEAT® was put on the market.

Table 1. Weights of values in cooking equipment

Value	1960	1965	1970	1972
Economy	27.8	27.8	24.3	22.9
Space	20.6	18.3	19.6	19.1
Time	13.2	14.1	13.8	14.1
Safety	15.5	13.6	13.7	13.7
Conformt	6.4	11.0	12.6	13.0
Convenience	16.5	15.2	16.0	17.2
Total	100.0	100.0	100.0	100.0

7. APPLICATION OF GRAPH THEORY AND MULTIDIMENTIONAL SCALING TO INTERACTIONS AMONG R & D ITEMS

7.1 Purpose

This paper describes a new method to visualize the relationship among technological development objects for supporting decision making.

In this method we combined graph theory with multidimensional scaling.

We demonstrated the application to the Japanese machinery and information industry.

The results of this study are as follows:

- We arranged many objects of development and presented figures easy to understand.
- We found a high efficiency group for investment in terms of technological relevance.
- We presented "the map of technology" which showed the relationship among objects in the high efficiency group.
- Developing new materials, new energy sources, and safety control technology for disasters are "key technologies".

7.2 Background

It was long ago that "a technology" existed independently in each field. it is common knowledge that the development of a certain technology in one field accelerates or decelerates technologies in other fields directly or indirectly. However, detecting the relationship among technologies and presenting it clearly is rather difficult. Nevertheless we must not avoid considering extensive development objects and technological relevance when planning a project.

When we plan a technological development strategy, we need an efficient method to arrange many proposed objects and to show clearly the interactions among them and the efficiency of investment. It is also important to present the result clearly and to make the planning speedy.

For this purpose we developed this method to treat a binary cross impact matrix of proposed objects.

In our method we employed graph theory and multidimensional scaling (hereafter MDS).

Our method is similar to usual cross impact matrix analysis for detecting interactions among technological objects, but it is much more widely applicable. The major distinctions between the cross impact matrix analysis and ours are as follows:

- The purpose of the former is forecasting, but that of the latter is supporting decision making.
- Because the former uses the Monte Carlo simulation, it needs a long computing time, which limits the number of objects. But since the latter doesn't use such simulation, it is more useful when handling a large number of objects.
- In the former method, numerical simulation is performed directly without considering the structure of interactions. However, we think it is better to present the structure clearly first. There are many cases where the presentation is quite enough to support decision making.

Generally, the graph theory in our method features as follows:

- We call each proposed object an "element".
- We think there are "interactions" among elements and that these don't change in a short time.

There are two ways to express such "elements" and "interactions". One is to consider points as "elements" and directed lines between each point as "interactions", namely, "a topological graph"; the other is "a matrix": row and column are considered and as "elements", each element of the matrix is considered as "interactions".

The great merit of the former is that it is visual and we can easily understand interactions at a glance. But in the case of many "elements" and "interactions", the graph becomes a "jungle" in which it is difficult to grasp the structure.

Although the latter isn't understandable visually and needs a large computer to operate large matrixes, the procedure is executed logically. These ways aren't so much opposed as complementary. In ours we operate the matrix by computer and present results by graph.

The MDS in our method is a procedure in which we express development objects by points and draw a map in multidimensional space - ordinary 2-dimension - from similarity between a pair of objects. Because we need a good map, all distances between each pair of points on the map must reflect optimally the degree of relation between developments. This is possible and is now a major development in psychometrics and statistics.

We tried to combine graph theory with MDS for operating a great number of interactions among development objects. This combination will be useful not only for comprehending the technological relevance of many development objects but also for deciding priority of the investment in technological development expenses, or for effectively planning an entry into a new market.

7.3 Method

Here we give an outline of the procedure.

(1) A 0-1 matrix of objects of technological development is an input datum, or for example in Fig. 1, in which we checked up all pairs of objects with respect to: "technology-A would affect positively technology-E directly, technology-B affects technology-C and F directly; C affects D; E affects B, F and G, and so on". In ours, "direct positive affection from X to Y" means "If we invest to develop X, we will expect a direct decrease of expense to develop Y or to speed up Y".

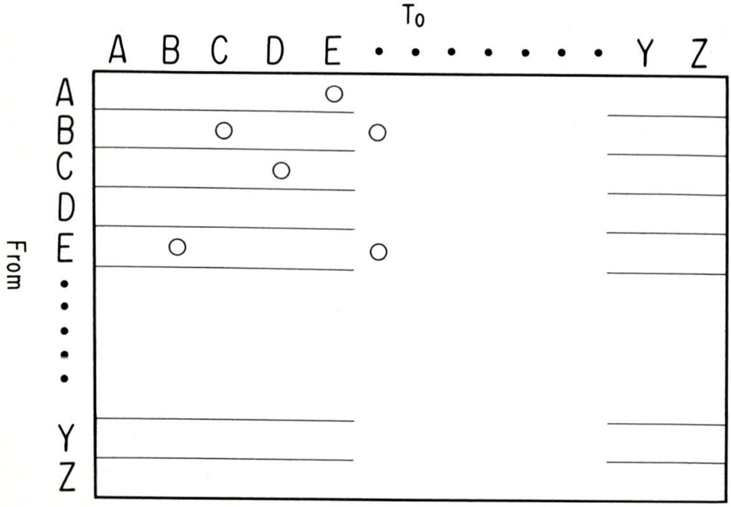

Fig. 1 A example of input datum

We can express it graphically "X ⟶ Y".
In Fig. 2, A,B,C,.....,K are objects of development. The arrow lines mean A affects positively only E, B affects positively C and F, etc.

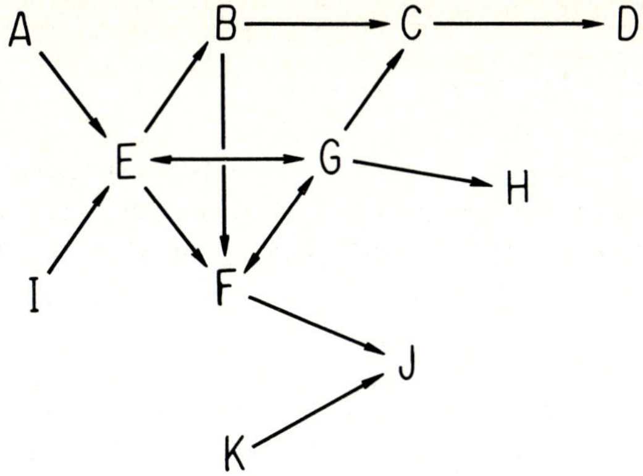

Fig. 2 A example of a topological graph

(2) Objects which neither affect nor are affected are omitted; those are called "independent technolgoies". But it should be noticed that sometimes an independent object leads to technological innovation if it succeeds.

(3) The object which affects much (i.e. many arrow lines go out) or is much affected (i.e. directed lines go in) is called "key technology".

(4) Strong component & condensed graph

We mentioned "direct affection" in section (1). Here we define an indirect affection. For example in Fig. 2, A affects B through E. A also affects C through E and B, through E and G, through E,F and G, for example. Thus, the relations between objects are also considered indirect effects.

Suppose we invest in G, and if we are successful in developing G, we will expect a positive effect on C,E,F and H directly, and on B,D and J indirectly, but not on A,I and

K, as Fig. 2 shows. This thus indicates the flow of effects.

Thus it is very helpful that we are able to arrange clearly the affect-affected chains among all points. Examining Fig. 2, we find any pair betweenn B,E,F and G is reachable directly or indirectly.

Let's make B,E,F,G one group. (Fig. 3)

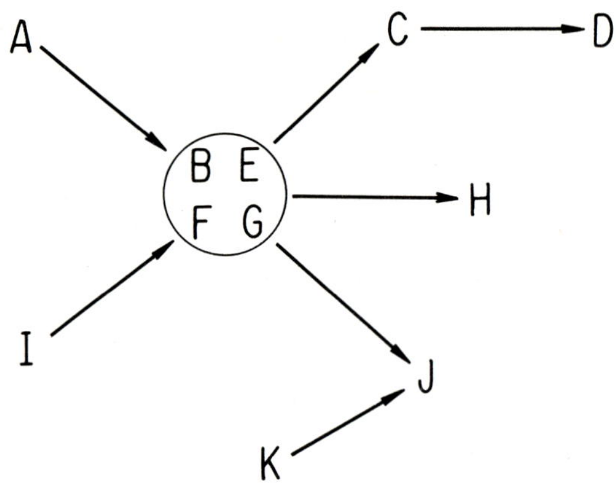

Fig. 3 Condensed graph

In Fig. 3, we can know, at a glance, what is affected positively when one object is developed, which suggests what an effective investment would be.

The "reachable-one-another" group (for example B,E,F,G) is called a "strong component", and a graph such as in Fig. 3 is called a "condensed graph", distinguished from the original graph shown in Fig. 2.

Because this example includes only 11 objects, we can draw it by hand. However, if the number of objects is 200 or 300, this will be impossible without using a large computer.

(5) A map of technology by multidimensional scaling

We will define "n-paths" and "length" between objects of technological development. In Fig. 4, when we pay attention to L and P, for example, there are ways like L → N → O → P and L → N → M → O → P. The former passes along three paths of arrow lines, the latter along four. So we call them "3-paths" and "4-paths", respectively. The minimum number of each n-paths between a point to a point is defined as the "length". In the above case the length is three.

The length doesn't satisfy the mathematical axiom of distance. So we define the "journey" as follows;

- The journey occurs only in the strong component.
- The journey is defined by the length; we think all arrow lines exist as bidirectional lines. For example, in Fig. 4, the journey betwenen L and P is 3, N and Q is 2, O and P is 1, and so on.

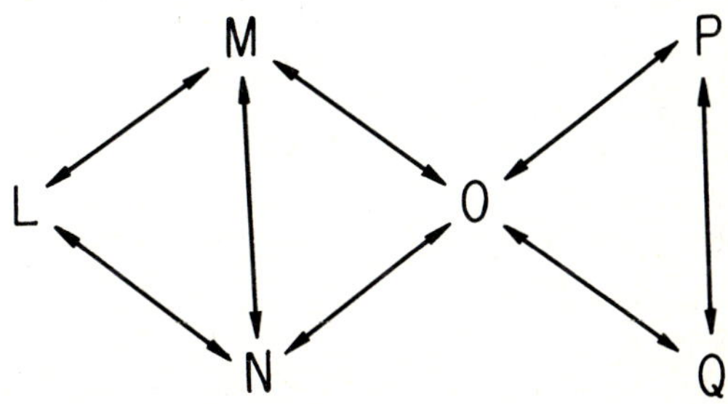

Fig. 4 A example of a graph

The journey satisfies the axiom.

Thus, when we take any strong component of a graph, we can measure the degree of effect between all pairs of objects of development by the journey. But in contrast with this, when we are given the information of the degree of affection among objects by the journey, it is difficult to make a map by Euclidian distance on which we distribute points to the optimal positions that reflect these journeys.

As mentioned above, now it is possible by MDS. Though it can't be necessarily plotted on a plane, we make an effort to construct a distribution which expresses a given journey information at maximum on 2-dimensional space. For this purpose many algebraic solutions and algorisms have been proposed in statistical and psychometrical fields, for example Theorem of Young & Householder (1938), Theorem of Schonemann (1972), TORSCA (Torgerson 1958), MDSCAL (Kruscal 1964), SSA (Guttman 1965), MDA (Hayashi 1974), etc.

In our method we adopt ADDSCAL, which is MDS for interval dissimilarities by estimating the optimal additive constant. (Yanai & Takane 1977, Takane 1980)

(6) The following information is also output;

- Which object of 2-paths and 3-paths downstream are affected by one object?
- Which objects of 2-paths and 3-paths upstream affect one object?
- What "length" can one object have to reach others?

7.4 Application to the Japanese Machinery and Information Industry

(1) Data

We applied this method to objects of technological development in "the table of elemental technology classified by technical field in the machinery & information industry" in "The Report for Strengthening the International Competing Power of the Japanese Machinery & Information Industry(1980)".

Objects for analysis are as follows:

Energy	55 items
Saving Energy	30 items

Natural Resources	26 items
Mateirals	29 items
Processing, Assembling	36 items
Living	32 items
Information	49 items
Others	36 items

For a total 293 items, we examined all interactions among them (293x293), and made a binary cross impact matrix, in which row-I column-J has a value of 1 if object-I affects J directly and positively, and for others the value is 0.

Table 1. Key technology

Items which have a large effect	A number of 1-path	2-path	3-path
Inorganic materials for heat	29	195	1134
Information system for anti-pollution	18	127	1032
Collection and transportation of waste	18	116	1015
Sensor	17	122	742
Monitoring system	15	103	742
Earthquake & fire-resistant architecture	15	113	838
High temperature gas furnace	14	119	795
Advanced boring	14	90	584
Community infomation system	13	96	797
Synthetic communication system	13	92	772
Remote maintenance	13	73	545
Information system for prevention of crime and accident	12	112	839
Corrosion resistant materials	12	60	469
Disposal of coal ash	12	47	326
Advanced thermal reactor	12	103	693
Calamities prediction	12	81	700
Data bank system	12	85	759
Inorganic function materials for chemistry	11	60	300
Energy complex	11	89	707
Fast breeder reactor	11	99	662
Safety engineering for nuclear reactor	11	82	665
High-strength materials	11	68	523
New bonding technique	11	76	414
Automatic self-diagnosis	10	75	609
Waste-heat boiler	10	72	490
Reprocessiong of spent fuel	10	69	536
Oil shale	10	72	477
Collecting sea uranium	10	40	280
High-temperature gas-turbine	10	74	438
Low-temperature burbine	10	73	524
Low-cost pattern recognition	10	62	434
Inorganic functional materials for optics	9	34	224
Heat-resistant materials	9	50	417
Disposal of radioactive waste	9	64	493
Oil sand	9	64	435
New technique of coal mining	9	60	368
Fatigue-resistant materials	9	34	267
Picture data digital communication	8	58	452
Ocean energy	8	56	321

Table 1 (continued)

Items which have a large effect	A number of 1-path	2-path	3-path
Corrosion resistant materials	38	194	1765
High-strength materials	33	151	1342
Heat-resistant materials	28	138	1154
Inorganic opt materials	25	109	808
Assessment method	24	130	874
Collection and transportation of waste	20	114	974
Information system for anti-pollution	20	119	960
Disposal of radioactive waste	17	129	899
Large area guard for preventing disaster	16	106	837
Collecting sea uranium	14	103	831
Reprocessing of spent fuel	14	123	866
Community information system	13	80	605
Picture data digital communication	13	106	712
Oceal thermal energy conversion	13	86	675
Flexible robot	13	68	524
Integrated communication system	13	80	591
Processing of waste gas	12	94	649
High-temperature gas furnace	12	94	649
Fatigue-resistant materials	11	49	524
Direct combustion of coal	11	87	526
Uranium survey	11	53	336
Earthquake & fire-resistant architecture	10	85	644
Energy complex	10	53	444
Fast breeder reactor	10	60	425
Coal gasification	10	49	381
Coal complex	10	68	462
Advanced thermal reactor	10	60	425
Excavation of abyssal floor	10	80	481
Information system for traffice & physical distribution	10	91	632
Heat source survey	9	48	298
Uranium enrichment	9	71	498
Cracking of heavy-oil	9	43	286
Electric power generation by coal gasification	9	71	462
Safety engineering for nuclear reactor	9	63	378
Inorganic materials for selective adsorbent	8	63	466
Large area airconditioning	8	49	283
Satelite for resource survey	8	51	239
Materials for low-temperature	8	36	205
Integrated traffic control system	8	54	409
VLSI	8	12	62

(2) Table 1 shows high ranking technologies of 1-path (40 items) which have a large effect and are affected by many. Values in Table 1 are numbers of 1-path, 2-path and 3-path from or to the point (technology).

We've shown clearly that most parts of Table 1 are developing new materials, new energy sources, and safety control technology to prevent disasters.

Software technology has not ranked high except for public systems.

(3) A condensed graph & strong components

The longest journey is 11 in our 293 × 293 binary cross impact matrix.

The 293 objects condensed as follows (Fig. 5: Condensed Graph):

Independent (having no line)	52 items
Included in a maximum strong component (hereafter MSC)	173 items
Included in strong components (5 components each with 2 objects)	10 items
Others	55 items

As Fig. 5 shows, the structure of interactions is a star-like shape and it is classified into three large groups as in the following:

* Group affecting MSC (hereafter Group-1)
* Group affected by MSC (hereafter Group-2)
* Group equal to MSC itself (hereafter Group-3)

As we show in Fig. 5, Group-1 mainly locates on the left hand side and Group-2 on the right hand side; we can thus easily see which object is included in each group.

In planning a strategy of technological development, Fig. 5 would suggest the following:

* If we invest in some of Group-1 and develop successfully, Group-3 or Group-2 would be developed at less expense. But no other objects in Group-1 would expect the economy. The effect on Group-2 would be indirect.
* Suppose we invest in an object in group-3; we would expect a positive effect on others in Group-3 and would expend less for developing any object in Group-2. It would also not be expected to try to affect Group-1. Since any pair of objects in Group-3 affects each other object directly or idirectly by definition, it is more efficient to concentrate on this group than to invest in Group-1 or Group-2.
* Because Group-2 is passive, investing in it first would be less efficient than investing in other groups. But it must not be forgotten that there are many cases in which we should invest because of a social demand or a business policy regardless of such inefficiency.

(4) A map of technology by multidimensional scaling

Because we need more detailed information for MSC (Group-3), we apply MDS which makes an Euclidian space map to express the degree of indirectness between each pair of objects by journeys. (Fig. 6)

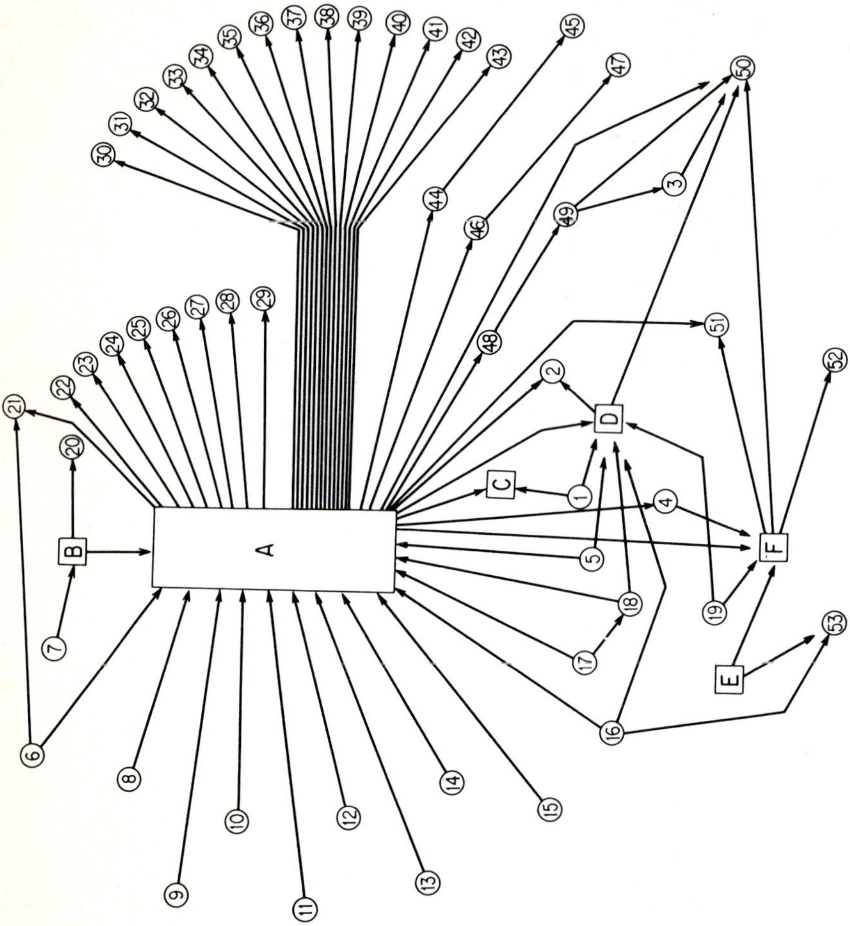
A
B
C
D
E
F
1
2
3
4
5
6
7
8
9
10
11
12
13
14
15
16
17
18
19
20
21
22
23
24
25
26
27
28
29
30
31
32
33
34
35
36
37
38
39
40
41
42
43
44
45
46
47
48
49
50
51
52
53

Fig. 5

A Maximum strong component (173 items)
B Satellite relay systems
Communication satelites
C Speech recognition
Voice code translation
D Sensor
Advanced pattern recognition
1 Inorganic functional materials for sound systems
2 Low-cost pattern recognition
3 Integrated CAM systems
4 Materials for lasers
5 Inorganic functional materials for electromagnetic products
6 Utilization of waste materials
7 Space information gathering
8 Reprocessing of waste oil
9 Inorganic functional materials for transportation
10 Elements to transformations energy to signals
11 Advanced logic devices
12 Flexible gripping
13 Super-skyscrapers
14 Gigantic structures
15 Keeping freshness of foods in distribution
16 Very accurate positioning
17 Artificial organs
18 Inorganic materials for organisms
19 Tracking systems
20 Broadcast satellites
21 Sewage reprocessing
22 Home refuse disposal
23 Reprocessing of metals
24 Low-temperature & low-pressure synthesis
25 Adiabatic engineering for productive process
26 Inorganic functional materials for dynamics
27 Adaptive control systems
28 Intelligent I/O units
29 Data bank systems
30 Nuclear fusion
31 Electric power generation by hydrogen
32 Pumped storage power stations
33 Reprocessing of waste water
34 Interactive systems for automatic design
35 Advanced operater console display
36 Non-impact printer
37 Associative memory
38 Local network
39 Handwriting reading machines
40 Rockets
41 Utilization of space energy
42 Aircraft engines
43 Planning, forecasting & evaluation engineering
44 Explication & control or photosynthesis
45 Industrial use of photosynthesis
46 Kanji reading machine
47 Kanji input units
48 Learning systems
49 Automatic work analysing
50 Monitoring systems
51 New bonding technique
52 Advanced opto device

FIG. 5 Condensed graph

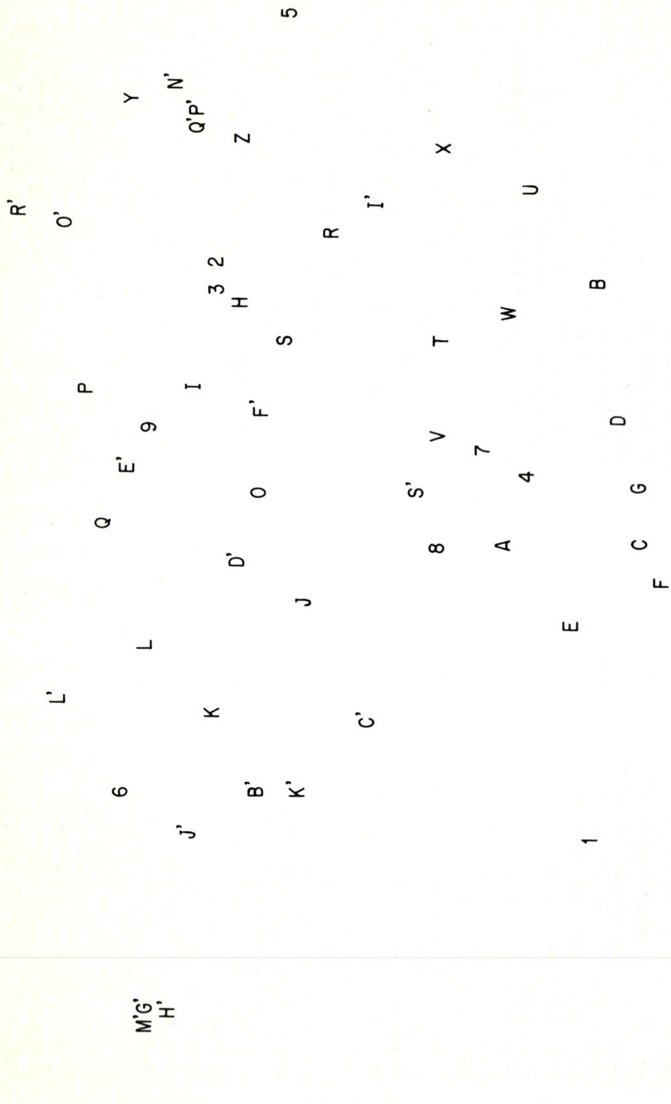

Fig. 6

1	Pyrolysis of waste plastics
2	Corrosion resistant materials
3	High-strength materials
4	Heat-resistant materials
5	Fatigue-resistant materials
6	Inorganic functional materials
7	Inorganic functional materials for heat
8	Inorganic materials for chemistry
9	Flexible robot
A	Oil shale
B	Coal gasification
C	Direct combustion of coal
D	Electric power generation by coal gasification
E	Processing of waste gas
F	Disposal of coal ash
G	Coal complex
H	Collecting sea uranium
I	Uranium survey
J	Uranium enrichment
K	Reprocessing of spent fuel
L	Disposal of rediactive waste
O	Advanced thermal reactor
	Fast breeder reactor
	High-temperature gas furnace
P	Safety engineering for nuclear reactor
Q	Nuclear power ships

R	Ocean thermal energy conversion
S	Advanced boring
T	High-temperature gas-turbine
U	MHD generators
V	Energy complex
W	Waste-heat boiler
X	Low-themperature turbine
Y	Diving dredgers
Z	Excavation of abyssal floor
B'	Information system for physical distribution
C'	Collection and transportation of waste
D'	Large area guard for preventing disaster
E'	Calamities prediction
	Automatic self-diagnosis
F'	Earthquake & fire - resistant architectures
G'	picture data digital communication
H'	Integrated communication system
I'	Remote maintenance
J'	Information system for traffic
K'	Information system for anti-pollution
L'	Information system for prevention of crime and accident
M'	Community information system
N'	Deep-sea Oil-drilling
O'	Structures in the sea
P'	Ocean energy
Q'	Ocean culture

FIG. 6 Multidimensional scaling

Displaying all 173 objects in group-3 on a map would be so hard to see that we show only the main 57 objects, which we choose as the logical sum of the six criteria as follows:

* 35 most affecting objects with 1-path. (having many directed lines from there).
* 35 most affecting objects with 2-paths.
* 35 most affecting objects with 3-paths.
* 35 most affected objects with 1-path. (having many directed lines to them).
* 35 most affected objects with 2-paths.
* 35 most affected objects with 3-paths.

Dissimilarity data for MDS is the journey between each pair.

The following should be noticed in this map:

- All points on the map can be reached.
- Generally, the nearer the two points, the stronger the affection between them.

7.4 Discussion

Thus we could arrange rationally many objects of development which are examined by each pair. We could present the result in figures easy to understand and could get to the heart of interactions of a large binary cross impact matrix at a glance. Moreover, we could find the most efficient group for investment with respect to its interactions. In the highly efficient group we could also derive a map of technology from which we could know the degree of affection among development objects from one point of view.

In the application to the Japanese machinery & information industry we've found that developing new materials, new energy sources, and safety control technology to prevent disasters were "key technologies".

This method presents strong support to allocate available resources or to decide a strategy for an entry into new technological fields.

But this method may be improved in the following ways:

- It must be considered how to make input data objective and rational.
- Although in this paper interactions among technological developments were expressed by 0-1 binary, we hope to be able to handle a large matrix which expresses the degree of relaitonship between a pair of objects, for example, on a 4-point rating scale. The descriptive name of those categories might be: very influential, moderately influential, somewhat influential, non influential.
- It will be very useful for decision making to store the output of this method in a data base for information retrieval by a time sharing system terminal; i.e. a sort of office automation.

APPENDIX 1: BIRTH OF SOFT SCIENCE IN JAPAN

Soft Science Study Committee was organized by the Science and Technology Agency of the Japanese Government in 1970. It consisted of Mr. Masao Sugimoto (Hitachi Ltd.), Mr. Takemochi Ishii (Tokyo Univ.), Mr. Zenji Katakata (Tokyo Institute of Technology), Mr. Reikichi Shirane (Nippon Telegram and Telephone), Mr. Hiroshi Nakai (Japanese Information Center for Science and Technology), Mr. Makoto Matsudaira (Rikkyo Univ.) and Mr. Tohru Yoshimura (Saitama Univ.). It presented a report after fifteen meetings in 1971. The outline of the report is as follows.

1. Expectation for Soft Science

   Qualitative change is occuring due to increasing importance of software in computer science, expanding application fields of systems engineering etc. Moreover, it became important task to take in soft factors such as social, human, spritual etc. in developing new technologies.

On the other hand, sign of basic change appeared in social and economic aspects for instance shortage of intellectual labor, complication of social phenomena, increase of changing speed, informationalization of industry and change of the value system. These are going to form roots of various problems. And request for new science and technology to cope with these changes.

Recently social engineering, operations research, systems engineering, behavior science, social ecology etc. are being developed in close connection with information science. These are soft science and technology with the possibility to cope with the above mentioned new request. And their future development is expected but they are not independent of mother's influence yet and far from full application to solve many political problems now we are facing.

Therefore, it is necessary to form a domain concept of soft science adopting these new science and technology and to extract important fields and to foster them positively.

2. How to Research and Develop Soft Science

(1) Target for research and develop

Among the important political problems now we are facing the role of soft science is especially expected to solve problems related with social evelopment such as public hazard, environmental pollution, urban problem, disaster prevention, transportation, education, medical care, crime prevention, public peace and order etc.

Approach process where the role of soft science is expected in solving these problem is forecasting of new problem, relevance analysis from total viewpoint, integrated planning, system design from total viewpoint, systemic creation, feedback information system, evaluation of planning and system, clarification of the value system etc. Especially, forecasting is essential and indispensable to solve various political problems.

Techniques and knowledge necessary for the approach is not sufficient yet, but the following fields are especially important and expected to fullfil.

As for techniques followings are necessary: forecasting technique, evaluating technique, planning technique, system design, system analysis, relevant analysis, behavior analysis, value analysis, simulation, gaming, creative technique, Delphi method, ecological approach, man-machine system, information system, etc.

Meanwhile, as for basic areas followings are important: information science, behavior science, mathematical science, social ecology, forecasting theory, computer science, creative engineering, social engineering, intellectual activity (judgement, recognition, learning etc.).

(2) Basic consideration for research and development promotion

Soft science is basic science and technology for solving public policy problem and it is essential for increasing national welfare and developing national economy, therefore, the role of government to promote it is important.

Most of research results of soft science nowadays were accumulated in the United States during approach process for actual problem with many years and much public funds. But in Japan it is desirable to promote research and development aiming directly and totally to level up soft science independent of the U.S. considering Japanese conditions are different from the U.S.'s.

Besides, analysis of issues and planning of measures should be done with care for special features different from hard science and technology in developing soft science.

3. The Status of the Art of Soft Science and Issue

Large-scale profit and non-profit research organizations exist in the U.S. which are soft science oriented such as Rand Corporation. And they are steadily obtaining excellent results.

Recently, so-called think-tank laboratories were established by economic groups in Japan. But it is difficult to tell that they are organization to research soft science comprehensively and their level is extremely low. Most important issues among these circumstances are, (1) Lack of interdiciplinary research system and mobility of research staffs, (2) Lack of flexibility of research organization, (3) Lower evaluation for intellectual technology such as information and software.

Besides, very few people come in to this field due to the vices of education system of Japan and conservative and feudalistic environment and climate.

4. Basic Concept for Comprehensive Research and Development System

As above mentioned, the status of the art of soft science is very backward in research system and training and maintaining personnels and new concept is requested to be developed for (1) establishing comprehensive research and development network, (2) establishing comprehensive research and development organization, (3) preparation of selecting and evaluating organization for research and development projects, (4) establishing personnel training system.

But in realizing them, there expected various obstacles of institution, operation and finance. It is necessary to make effort to realize the above mentioned from the view point of long term, and to start one by one which can be started now.

5. Measures to Develop Soft Science

(1) Preparation of comprehensive research and development system

It is necessary to form comprehensive research and development network through universities, national and public test and research organizations, private research organizations and then to establish comprehensive research and developoment organization in the nucleus of the network. This organization should have such function as research and development, personnel training, clearing and data bank and research results of soft science should be accumulated. It is also needed to assemble comprehensive research and development projects from various individual problems brought up by industries and administrative organizations. It is also necessary to settle a function to evaluate research results obtained by research organizations.

(2) Increase of governmental contract research project in soft science

It is necessary to stimulate research and development in governmental organizations to level up soft science through approach to actual projects as well as to increase orders of contract research for private research organizations.

(3) Training of personnels

It is necessary to establish interdisciplinary research system and to train personnel in it, as well as to adopt system or education method to train specialists in interdisciplinary field in research organizations like universities.

(4) Promotion of human mobility

In order to promote interdisciplinary research system it is necessary to utilize excellent people actively. For this purpose it is necessary to improve various systems for specialists to move among universtities, national and public laboratories and private laboratories.

(5) Preparation of information distribution system

It is necessary to prepare clearing function and data bank needed to promote research and development in soft science.

(6) Aids for soft science research

As soft science is basic science and technology playing big role in solving polilitical problem, it is necessary to increase research and development budget for this field in governmental laboratories as well as in private laboratories.

APPENDIX 2: SIMILAR CONCEPT WITH SOFT SCIENCE

1. Comprehensive Research and Development

Basic, applied and developing research study incorporating various knowledge of specialities such as economy, society, technology etc.

2. Policy Science

Interdisciplinary science related with various phenomena in policy determination and execution.

3. Social Engineering

Academic field to study contact area of science and technology, and, human and social science. It is used to judge project comprehensively by integrating knowledge of science, technology, society and human being.

4. Applied Systems Analysis

Research strategy that involves the use of techniques, concepts and scientific, systematic approach to the solution of complex problems.

5. Technology Assessment

Every technological development has positive and negative influences for the society. This is an approach to assess the influence of the development before it is developed.

ACKNOWLEDGEMENT

The author thanks Messrs. Naoharu Miyakawa, Taro Tanimatsu, Tetsuo Kobayashi, Koju Ueda for their active participation in preparing this chapter. Each section of this chapter wholly owes to their works and the author merely edited their works.